T0184690

Teaching Science in Out-of-School Settings

Junqing Zhai

Teaching Science
in Out-of-School Settings

Pedagogies for Effective Learning

 Springer

Junqing Zhai
College of Education
Zhejiang University
Hangzhou, China

ISBN 978-981-10-1250-1   ISBN 978-981-287-591-4   (eBook)
DOI 10.1007/978-981-287-591-4

Springer Singapore Heidelberg New York Dordrecht London
© Springer Science+Business Media Singapore 2015
Softcover reprint of the hardcover 1st edition 2015
This work is subject to copyright. All rights are reserved by the Publisher, whether the whole or part of
the material is concerned, specifically the rights of translation, reprinting, reuse of illustrations, recitation,
broadcasting, reproduction on microfilms or in any other physical way, and transmission or information
storage and retrieval, electronic adaptation, computer software, or by similar or dissimilar methodology
now known or hereafter developed.
The use of general descriptive names, registered names, trademarks, service marks, etc. in this publication
does not imply, even in the absence of a specific statement, that such names are exempt from the relevant
protective laws and regulations and therefore free for general use.
The publisher, the authors and the editors are safe to assume that the advice and information in this book
are believed to be true and accurate at the date of publication. Neither the publisher nor the authors or the
editors give a warranty, express or implied, with respect to the material contained herein or for any errors
or omissions that may have been made.

Printed on acid-free paper

Springer Science+Business Media Singapore Pte Ltd. is part of Springer Science+Business Media
(www.springer.com)

For my parents with thanks

Foreword

It is my great pleasure to write this Foreword. Back in 2007 when Junqing applied to carry out his PhD studies with us at King's College London (where I was working at the time), he was proposing to look at aspects of environmental education in his home country, China. One of his research questions was 'What are the problems or challenges in practical implementation of environmental education in Chinese educational system?' It soon became clear after he joined us in London that his interests had evolved, and his question by the end of the year was 'What might outdoor experiences contribute to young children's learning?' As he started to read more and visit a number of environmental education practitioners, Junqing's interest shifted further and he became increasingly focused on the role of botanic garden educators in the education of school visitors—the topic of this volume.

What you have here is a really thorough piece of scholarship carried out intelligently, painstakingly and with passion. It almost goes without saying that it wasn't always easy, but we had an outstanding research group who supported each other through the bad times and celebrated the good times.

As Junqing says himself, 'the strength of this study lies in the production of an in-depth, contextual understanding of botanic garden educators' pedagogical thinking and choices that engage and support visiting schoolchildren's learning of ecological science'. The study fills a gap in the literature and brings together ideas about science education pedagogy—particularly questioning—and the nonschool context of botanic gardens. The study covers new ground in terms of the focus of the study as well as introducing some innovative methodological approaches. It is also important to point out that the findings of the research have practical applications not just in England, where the study was carried out, but in any country where school students are educated about ecological science in botanic gardens and other natural surroundings.

Unusually, Junqing's thesis needed no corrections after he had his viva. I had the privilege of sitting in on the oral examination, and the end result was never in doubt. But it takes a lot of work to turn PhD thesis into an academic book, and Junqing has laboured long and hard to produce this volume. It is well-written, appropriately academic and very readable. I hope that it stimulates your thinking and, if you are a doctoral student, it gives you some inspiration.

University of Bristol, Bristol, UK   Justin Dillon

Acknowledgement

The publication of this book was supported by the Zhejiang University Gong Haoran Fund for Research and Publication on Vygotsky.

Contents

Chapter 1
Introduction

Public gardens, parks and botanic gardens attract a large number of international and domestic visitors throughout the world. According to the statistics of the Botanic Gardens Conservation International (BGCI), there are over 2,000 botanic gardens in the world with around 250 million visitors each year. As one of the most popular informal settings for organized school visits, botanic gardens are 'living museums' of plants where students acquire practical biological knowledge, develop horticultural skills, learn to appreciate the natural environment and develop a sense of sustainability (Braund & Reiss, 2006). In other words, they play a significant role in providing the public with an enjoyable way to spend leisure time as well as giving them the opportunity to explore the pleasures of the garden environment, whilst also serving as a site for enhancing public education (Johnson, 2004; Sanders, 2007; Stewart, 2003; Tunnicliffe, 2001). Recent research has indicated that school educational excursions to informal settings have a positive impact on students' cognitive learning, affective maturation and social development (DeWitt & Storksdieck, 2008; Rennie, 2007). Every year, in the United Kingdom, each of the 130 or more botanic gardens throughout the country accommodates a substantial number of visiting school groups to support their study of plants and many of these are guided by professional botanic garden educators (BGEs).

The most common reason for teachers taking their students to informal science institutions is to enhance ideas within the curriculum or to provide enrichment beyond it, thus expecting students to learn content and increase their motivation to study science (Kisiel, 2005; Tal & Steiner, 2006). However, a growing number of studies have shown that learning opportunities during school visits to informal settings can be missed because educators tend to attend to tasks that have to be completed or explanations that need to be covered (DeWitt & Osborne, 2007; Tal & Morag, 2007). Moreover, it was found that many schoolteachers lack the confidence, competence and expertise to teach students outside of the classroom (Glackin, 2007; O'Donnell, Morris, & Wilson, 2006). Regarding this point, a recent study of school visits to a museum reported that teachers prefer guided visits to non-guided ones as they do not have as good a mastery of the themes under consideration as the museum

© Springer Science+Business Media Singapore 2015
J. Zhai, *Teaching Science in Out-of-School Settings*,
DOI 10.1007/978-981-287-591-4_1

educators do (Faria & Chagas, 2013). Similarly, for school visits to botanic gardens, teachers often do not have intensive content knowledge about plants and resources available on site. Consequently, attending BGE-guided visits has become a common practice for visiting school groups although little is known about the pedagogical practices of this group of educators.

Ohlsson (1996) argues that learning to understand abstract knowledge, such as concepts, ideas and principles, requires learners to take part in a list of epistemic tasks, including: describing, explaining, predicting, arguing, critiquing, explicating and defining. That is, he pertained that it is essential for learners to be socially and intellectually engaged, if effective learning is to take place. In the same vein, according to Newmann et al. (1992), engagement is the doorway towards learning, understanding and mastering of knowledge and skills. Moreover, abundant research has shown that young people can benefit greatly from their experiences of learning outside the classroom (Kahn & Kellert, 2002; Nundy, 1999; Rickinson et al., 2004, for a review; South, 1999). However, little is known regarding whether they are actually engaged in learning during their educational visits to these informal settings and further, how the educators engage them with the different activities available in these settings (DeWitt, 2007; Dillon et al., 2005).

Gombert (1992) argues that 'the acquisition and restructuring of knowledge generally requires the conscious participation of the subject' (p. 195). In this regard, the conscious participation of the subject is the state of engagement in social activities, which is regarded as the final stage of knowledge construction in the individual's mind (Valsiner, 1997; van Lier, 1996; Vygotsky, 1978; Wertsch, 1991). In his summary of children's development, Vygotsky (1986) proposes that 'any function in the child's cultural development appears twice, or on two planes; first it appears on the social plane and then on the psychological plane' (p. 163). That is, a child acquires learning, first, through social interaction and this engagement provides the link to the psychological level, where the knowledge is internalized. Consequently, there is the need for education researchers to understand learning engagement, especially the strategies to engage people. Given that a number of scholars (e.g. Alexander, 2006; Brown, Collins, & Duguid, 1989; Edwards & Mercer, 1987; Greeno, Collins, & Resnick, 1996; Greeno & The Middle School Mathematics Through Applications Project Group, 1998; Mercer & Littleton, 2007; Mortimer & Scott, 2003; Ohlsson, 1996; Osborne, Erduran, & Simmon, 2004; Rogoff, 1998; van Lier, 1996; Vosniadou, 2001) have contended that learning starts from social interaction and that language should be considered as the most important tool for learning, then it follows that this forms one of the most significant resources for engaging learners in the process of knowledge construction. With respect to science education, Lemke (1990) argues that learning science involves learning to talk science through classroom discourse and terms this 'doing science through the medium of language' (p. xi). Furthermore, Mortimer and Scott (2003) have explained that 'different kinds of interactions between teacher and students in the science classroom contribute to meaning making and learning' (p. back cover).

According to previous studies carried out in science classrooms, one of the most efficient approaches for engaging children in making meaning out of scientific views is through talk (e.g. Duschl & Osborne, 2002; Lemke, 1990; Scottt, Mortimer, & Aguiar, 2006; Scott & Ametller, 2007) and the same claim has been made with regard to informal settings (e.g. DeWitt, 2007; Rahm, 1998; Rowe, 2002; Tunnicliffe, 2001). More specifically, according to Bruner (1996), young people need to think for themselves before they achieve conceptual understanding and teaching should provide them with those linguistic opportunities and encounters that will enable them to achieve this. Building on this perspective, in *TowardsDialogicTeaching: Rethinking Classroom Talk*, Alexander (2006, p. 10) posed the question 'Do we provide and promote the right kind of talk?' to guide classroom teachers to reflect on their teaching. Other researchers (Cazden, 2001; Edwards & Westgate, 1994) have pointed out that owing to it being an important pedagogical component, more concentration than hitherto should be paid to teachers' talk, especially in relation to how they engage their students in discourse. Moreover, DeWitt (2007) has called for future research on learning in informal contexts to focus on the nature of spoken practitioner exchange with students during visits.

In light of the existing literature that states that the educator plays a critical role in determining the success of a school trip as a learning experience (Anderson, Kisiel, & Storksdieck, 2006; Cox-Petersen, Marsh, Kisiel, & Melber, 2003; Griffin & Symington, 1997; Tal & Morag, 2007), the aim of this study is to explore the BGEs' role in supporting and enriching such a learning experience. In seeking to characterize the pedagogical practices of the BGEs, three questions are proposed to guide the investigation:

(a) What is the structure of the BGE-guided school visits?
(b) How do the BGEs interact with students in terms of engaging and supporting learning?
(c) What kind of pedagogical behaviours that facilitate students' learning can be observed during the guided visits?

In order to respond to these questions, a naturalistic case study enquiry was conducted, with the BGEs' practices in supporting and enriching children's visiting experiences being examined in detail. Data, including audio/video recordings of BGEs and students engaged in talk during the visits, field notes, interviews and student surveys, were collected for this study. The interviews with the BGEs identified their views of teaching and learning in botanic gardens and gave them the opportunity to reflect on their teaching practices as observed for this study. The investigation of the discourse between BGEs and students during the teaching and learning processes occurring in botanic garden settings provided insights into the quality of the pedagogical practices. Finally, to triangulate with the observation data, students' visiting experiences were researched through applying a post-visit survey.

Much of the strength of this study lies in the production of an in-depth, contextual understanding of BGEs' pedagogical thinking and choices that engage and support visiting schoolchildren's learning of ecological science. This study bridges the gap in the literature regarding science education in informal contexts. More specifically, by placing an emphasis on the pedagogical behaviours of BGEs, this study seeks to develop a detailed understanding of the role of mediators in botanic gardens. The methodological considerations and the frameworks devised from the analysis in Chap. 4 are offered as suggestions how to proceed to other researchers who intend to investigate teaching and learning in similar settings, for example: obtaining access to participants, data collection strategies and the development of analytical schemes. In practical terms, the findings have implications regarding the improvement of BGEs' practice, the design of education programmes and the designing of effective professional development schemes. In addition, a contribution is made in the form of a framework for the development of pedagogical theory relating to outdoor settings.

References

Alexander, R. (2006). *Towards dialogic teaching: Rethinking classroom talk* (3rd ed.). York: Dialogos.

Anderson, D., Kisiel, J., & Storksdieck, M. (2006). Understanding teachers' perspectives on field trips: Discovering common ground in three countries. *Curator: The Museum Journal, 49*(3), 365–386.

Braund, M., & Reiss, M. (2006). Towards a more authentic science curriculum: The contribution of out-of-school learning. *International Journal of Science Education, 28*(12), 1373–1388.

Brown, J. S., Collins, A., & Duguid, P. (1989). Situated cognition and the culture of learning. *Educational Researcher, 18*(1), 32–42.

Cazden, C. B. (2001). *Classroom discourse: The language of teaching and learning* (2nd ed.). Portsmouth, NH: Greenwood Press.

Cox-Petersen, A. M., Marsh, D. D., Kisiel, J., & Melber, L. M. (2003). Investigation of guided school tours, student learning, and science reform recommendations at a museum of natural history. *Journal of Research in Science Teaching, 40*(2), 200–218.

DeWitt, J. (2007). *Supporting teachers on science-focused school trips: Towards an integrated framework of theory and practice* (Unpublished doctoral dissertation). King's College London, London, UK.

DeWitt, J., & Osborne, J. (2007). Supporting teachers on science-focused school trips: Towards an integrated framework of theory and practices. *International Journal of Science Education, 29*(6), 685–710.

DeWitt, J., & Storksdieck, M. (2008). A short review of school field trips: Key findings from the past and implications for the future. *Visitor Studies, 11*(2), 181–197.

Dillon, J., Morris, M., O'Donnell, L., Reid, A., Rickinson, M., & Scott, W. (2005). *Engaging and learning with the outdoors: The final report of the outdoor classroom in a rural context action research project*. London: National Foundation for Education Research.

Duschl, R. A., & Osborne, J. (2002). Supporting and promoting argumentation discourse in science education. *Studies in Science Education, 38*, 39–72.

Edwards, D., & Mercer, N. (1987). *Common knowledge: The development of understanding in the classroom*. London: Routledge.

Edwards, D., & Westgate, D. (1994). *Investigating classroom talk*. London: Falmer.

Faria, C., & Chagas, I. (2013). Investigating school-guided visits to an aquarium: What roles for science teachers? *International Journal of Science Education, Part B, 3*(2), 159–174.

Glackin, M. (2007). Using urban green space to teach science. *School Science Review, 89*(327), 1–8.

Greeno, J. G., Collins, A., & Resnick, L. B. (1996). Cognition and learning. In D. B. Berliner & R. C. Calfee (Eds.), *Handbook of educational psychology* (pp. 15–46). London: Prentice-Hall.

Greeno, J. G., & The Middle School Mathematics Through Applications Project Group. (1998). The situativity of knowing, learning, and research. *American Psychologist, 53*(1).

Griffin, J. M., & Symington, D. (1997). Moving from task-oriented to learning-oriented strategies on school excursions to museums. *Science Education, 81*(6), 763–779.

Johnson, S. (2004). Learning science at a botanic garden. In M. Braund & M. Reiss (Eds.), *Learning science outside the classroom* (pp. 75–93). London: Routledge.

Kahn, P. H., & Kellert, S. R. (Eds.). (2002). *Children and nature: Psychological, sociocultural, and evolutionary investigations.* Cambridge, MA: The MIT Press.

Kisiel, J. (2005). Understanding elementary teacher motivations for science fieldtrips. *Science Education, 89*(6), 936–955.

Lemke, J. L. (1990). *Talking science: Language, learning and values.* Norwood, NJ: Ablex Publishing.

Mercer, N., & Littleton, K. (2007). *Dialogue and the development of children's thinking: A sociocultural approach.* London: Routledge.

Mortimer, E. F., & Scott, P. H. (2003). *Meaning making in secondary science classroom.* Maidenhead, England: Open University Press.

Newmann, F. M., Wehlage, G. G., & Lamborn, S. D. (1992). The significance and sources of student engagement. In F. M. Newmann (Ed.), *Student engagement and achievement in American secondary school* (pp. 11–39). New York: Teachers College Press.

Nundy, S. (1999). The fieldwork effect: The role and impact of fieldwork in the upper primary school. *International Research in Geographical and Environmental Education, 8*(2), 190–198.

O'Connor, M. C., & Michaels, S. (1996). Shifting participant frameworks: Orchestrating thinking practices in group discussion. In D. Hicks (Ed.), *Discourse, learning and schools* (pp. 63–103). Cambridge: Cambridge University Press.

Ohlsson, S. (1996). Learning to do and learning to understand: A lesson and a challenge for cognitive modelling. In P. Reiman & H. Spada (Eds.), *Learning in humans and machines: Towards an interdisciplinary learning science* (pp. 37–62). Oxford, UK: Elsevier.

Osborne, J., Erduran, S., & Simmon, S. (2004). Enhancing the quality of argumentation in school science. *Journal of Research in Science Teaching, 41*(10), 994–1020.

Rahm, J. (1998). *Growing, harvesting, and marketing herbs: Ways of talk and thinking about science in a garden* (Unpublished doctoral dissertation). University of Colorado at Boulder, Boulder, CO., USA.

Rennie, L. (2007). Learning science outside of school. In S. Abell & N. G. Lederman (Eds.), *Handbook of research on science education* (pp. 125–167). Mahwah, NJ: Lawrence Erlbaum.

Rickinson, M., Dillon, J., Teamey, K., Morris, M., Choi, M. Y., Sanders, D., et al. (2004). *A review of research on outdoor learning.* London: National Foundation for Educational Research & King's College London.

Rogoff, B. (1998). Cognition as a collaborative process. In W. Damon, D. Kuhn, & R. Siegler (Eds.), *Handbook of child psychology* (5th ed., Vol. 2, pp. 679–744). New York: Wiley.

Rowe, S. M. (2002). *Activity and discourse in museums: A dialogic perspective on meaning making* (Unpublished doctoral dissertation). Washington University, Saint Louis, Missouri, USA.

Sanders, D. (2007). Making public the private life of plants: The contribution of informal learning environments. *International Journal of Science Education, 29*(10), 1209–1228.

Sawada, D., Piburn, M. D., Judson, E., Turley, J., Falconer, K., Benford, R., et al. (2002). Measuring reform practices in science and mathematics classrooms: The reformed teaching observation protocol. *School Science and Mathematics, 102*(6), 245–253.

Scott, P. H., & Ametller, J. (2007). Teaching science in a meaningful way: Striking a balance between 'opening up' and 'closing down' classroom talk. *School Science Review, 88*(324), 77–83.

Scott, P. H., Mortimer, E. F., & Aguiar, O. G. (2006). The tension between authoritative and dialogic discourse: A fundamental characteristic of meaning making interactions in high school science lessons. *Science Education, 90*(4), 605–631.

South, M. (1999). Can a botanic garden education visit increase children's environmental awareness? In L. A. Sutherland, T. K. Abraham, & J. Thomas (Eds.), *The power for change: Botanic gardens as centres of excellence in education for sustainability. Proceedings of the 4th international congress on education in botanic gardens* (pp. 68–76). Richmond, Surrey: Botanic Gardens Conservation International.

Stewart, K. M. (2003). *Learning in a botanic garden: The excursion experiences of school students and their teachers* (Unpublished doctoral dissertation). University of Sydney, Sydney, Australia.

Tal, T., & Morag, O. (2007). School visits to natural history museums: Teaching or enriching? *Journal of Research in Science Teaching, 44*(5), 747–769.

Tal, T., & Steiner, L. (2006). Patterns of teacher-museum staff relationships: School visits to the educational centre of a science museum. *Canadian Journal of Science, Mathematics, and Technology Education, 6*(1), 25–46.

Tunnicliffe, S. D. (2001). Talking about plants: Comments of primary school groups looking at plant exhibits in a botanical garden. *Journal of Biological Education, 36*(1), 27–34.

Valsiner, J. (1997). *Culture and the development of children's action: A theory of human development*. New York: Wiley.

van Lier, L. (1996). *Interaction in the language curriculum: Awareness, autonomy and authenticity*. London\New York: Longman.

Vosniadou, S. (2001). *How children learn. Educational Practices Series, 7*. Brussels: The International Academy of Education (IAE)\The International Bureau of Education (UNESCO).

Vygotsky, L. S. (1978). *Mind in society*. Cambridge, MA: Harvard University Press.

Vygotsky, L. S. (1986). *Thought and language*. Cambridge, MA: Harvard University Press.

Wertsch, J. V. (1991). *Voices of the mind: A sociocultural approach to mediated action*. London: Harvester Wheatsheaf.

Chapter 2
Botanic Gardens as Teaching and Learning Environments

Botanic gardens generally comprise walled gardens, in which are displayed a wide range of plants in various environments, appropriately labelled with botanical names. Usually, they have long-standing affiliations with scientific research organizations that are engaged in researching plant taxonomy and other aspects of botanical science. However, when they were initially established, their remit was not as complex as it is today, in that their role has been extended to encompass the challenge of holding documented collections of living plants for the purposes of: scientific research, conservation, display and education (BGCI, 2008a).

2.1 The Rise of Botanic Gardens

The origins of modern botanic gardens can be traced to the physic gardens, which concentrated their work on cultivating medical and aromatic plants (Rae, 1996) and they were first founded during the Italian Renaissance in the sixteenth century. Their function was purely the academic study of medicinal plants (Brockway, 1979), and by the seventeenth century, these medicinal gardens had spread to universities and apothecaries across Europe (BGCI, 2008b). In fact, botany was not a distinct discipline when the early physic gardens were founded in the sixteenth century, because the focus of their work was on developing descriptive adjuncts for medicinal plants. When it came to the late seventeenth and early eighteenth centuries, botanic gardens began to feature in and contribute to the development of botany as a scientific discipline (BGCI, 2008b).

Advances in shipbuilding and navigation allowed Western countries to sail the oceans and explore new territories in the eighteenth century, which, as Brockway (1979) argued, 'added appreciably to botanical collections and spurred a great interest in botany as a science' (p. 451). As the British Empire expanded, the colonial plantations needed seeds, crops and horticultural advice in order to obtain better yields.

© Springer Science+Business Media Singapore 2015
J. Zhai, *Teaching Science in Out-of-School Settings*,
DOI 10.1007/978-981-287-591-4_2

Although botanic gardens had a responsibility of serving such colonial botany requirements, their research and education functions were also important. For example, Joseph Hooker, the first official director at Kew Gardens, pursued scientific autonomy for the institute and listed its functions as: 'display and public education; collection and classification of plants; research, with a special laboratory for the study of plant physiology, cytology and genetics; publication; information storage and retrieval; and a training program ... by sending hundreds of botanists and gardeners to all the colonial gardens, to the universities and to the great commercial nurseries' (Brockway, 1979, p. 453).

As decay of the British Empire took hold and with the independence of the colonies during the twentieth century, botanic gardens no longer served their earlier role of addressing the demands of colonial rule. Instead, their conservation role became salient, particularly in the second half of the twentieth century, when there were growing concerns relating to climate change and the loss of biodiversity. For instance, the *World Conservation Strategy* (International Union for Conservation of Nature and Natural Resources, 1980) was one of the key environmental policy initiatives, which advocates conserving ecosystems and natural resources to provide for sustainable development. Further, the World Commission on Environment and Development's (WCED) seminal report *Our Common Future* (1987) addressed the interdependent nature of the relationship between the environment and development and the authors advocated a stance towards human development. This particular report points out that in order to achieve sustainable development, specific attention needs to be paid to the conservation of species and ecosystems, as they constitute the fundamental bases of development. The founding of the Botanic Gardens Conservation International (BGCI) in 1987 recons the development and implementation of global policies related to environmental protection. The BGCI's mission endeavours to 'ensure the world-wide conservation of threatened plants, the continued existence of which are intrinsically linked to global issues including poverty, human well-being and climate change' (BGCI, 2008c).

More recently, the *International Agenda for Botanic Gardens in Conservation* (Wyse-Jackson & Sutherland, 2000) sets out guiding principles for botanic gardens worldwide to promote plant conservation through research and education. To monitor the implementation of this agenda, the BGCI launched a guiding document—*2010Targets for Botanic Gardens*—which urges the leaders of botanic gardens worldwide to: (1) understand and document plant diversity, (2) conserve plant diversity, (3) use plant diversity sustainably, (4) promote education and awareness about plant diversity and (5) build capacity for the conservation of plant diversity. Moreover, the battle against the loss of biodiversity and other environmental problems continues to be a pressing issue during the twenty-first century, and for botanic gardens, in particular, further challenges will be encountered in the struggle to achieve plant conservation and sustainability.

2.2   Philosophies of Learning in Garden Settings

The natural environment has been considered as a robust educational site by many educationalists for centuries and school gardens and botanic gardens are no exception. A number of the most influential Western educational philosophers and pioneer thinkers, such as Comenius, Rousseau, Pestalozzi, Froebel, Montessori and Dewey, viewed gardens as significant educational settings (Sanders, 2004; Subramaniam, 2002). In this section, various theorists' comments on education in outdoor settings, especially in gardens, are reviewed.

The father of modern education, Czech educationist and philosopher Johann Comenius (1592–1670), characterized human life from the mother's womb to the grave as a series of educational stages, in which objects from nature could serve as the basis of learning (Comenius, 1660). He stated that 'education should be universal, optimistic, practical and innovative and that it should focus not only on school and family life but also on social life in general' (Rowe & Humphries, 2004, p. 19). Further, he argued that knowledge begins from sense passing into memory through imagination and only can then the understanding of universals be achieved (Boyd & King, 1995). Although Comenius's views on knowledge acquisition are close to materialist sensationalism, his principal belief that teaching and learning should follow a natural process still influences today's curriculum and pedagogy. According to him, seeing, hearing, tasting and touching are the key methods whereby children become acquainted with water, earth, air, fire, rain, stone, iron, plants and animals, which prepare the way for understanding the natural sciences. Consequently, he suggested that 'a school garden should be connected with every school where children can have the opportunity for leisurely gazing upon trees, flowers and herbs and are taught to appreciate them' (Rowe & Humphries, 2004, p. 19). As a response, Rowe and Humphries (2004) stated, 'Comenius's advocacy of an authentic curriculum led us to develop the outdoor setting as our largest classroom' (p. 19).

Jean-Jacques Rousseau (1712–1778), the French philosopher, believed that human beings were happy when in a state of nature, but were corrupted by society, and contend that nature is the best teacher for children. According to this naturalist point of view, education should 'focus on the environment, on the need to develop opportunities for new experiences and reflection and on the dynamic provided by each person's development' (Darling, 1994, p. 82). Johann Heinrich Pestalozzi (1746–1827) agreed with Rousseau's child-centred educational perspective and suggested that teaching should focus on observation and activity, rather than only on words. He put his educational thoughts into practice in Yverdon, Switzerland, by establishing a school to teach orphans gardening, farming and home skills. Although Pestalozzi's educational innovation failed as his school went bankrupt, his concept of achieving a balance between the three elements, hands, heart and head, still influences the field of education, seen for instance in contemporary commitment to providing authentic learning environments and worthwhile hands-on activities.

Fredrick Froebel (1782–1852), a student of Pestalozzi, believed that 'humans are essentially productive and creative and that fulfilment comes through developing these characteristics in harmony with God and the world' and through his work he tried to 'encourage the creation of educational environments that involved practical work and the direct use of materials' (Smith, 2008). Moreover, he viewed play as an important way of engaging children in learning, because it stimulates their interest, and his first kindergarten, established in 1840, was designed to promote children's awareness of the natural world through observing and nurturing plants in a play rather than a formal education setting. In short, Froebel emphasized doing as well as observing to motivate children to become involved in learning, and Sealy (2001) and Subramaniam (2002) concluded that he was one of the most effective proponents of school gardens in the nineteenth century.

Maria Montessori (1870–1952) similarly addressed the educational function of gardens and advocated an active engagement with them, rather than a contemplative one (Montessori, 1912; Sanders, 2008). She realized that children's gardens could be used beyond the standard curriculum to help to 'develop patience, enhance moral education, increase responsibility and improve appreciation for nature and relationship skills' (Montessori, 1912, pp. 156–160). John Dewey (1859–1952) criticized her methods, because she ignored the importance of the social interaction of participants, but both of them agreed that students should be at the centre of the whole process of education. Dewey (1938) emphasized the salience of the children's experience and argued that educators must first understand the nature of human experience. He argued that children should be involved in real-life tasks and challenges, such as outdoor excursions, weaving and construction in wood, and in particular, he noted the potential educational function of gardening. In *Democracy and Education*, Dewey (1916) highlighted the importance of gardening in a chapter entitled 'Play and Work in the Curriculum' as follows:

> Gardening need not be taught either for the sake of preparing future gardeners, or as an agreeable way of passing time. It affords an avenue of approach to knowledge of the place farming and horticulture has had in the history of the race and which they occupy in present social organization. ... Instead of the subject matter belonging to a peculiar study called botany, it will then belong to life and will find, moreover, its natural correlations with the facts of soil, animal life and human relations. As students grow mature, they will perceive problems of interest which may be pursued for the sake of discovery, independent of the original direct interest in gardening–problems connected with the germination and nutrition of plants, the reproduction of fruits, etc., thus making a transition to deliberate intellectual investigations. (pp. 163–164)

In the long historical period during which educational philosophy has emerged, gardens often have been considered as an important place for teaching and learning. The philosophies of the educationists reviewed above demonstrate a shared understanding of the role of education in appreciating and valuing nature; in other words, these theorists have claimed that children's experiences with the natural world can contribute to their individual development. However, this contention has been criticized for having weaknesses, such as it ignores children as members of civil society (Falk & Dierking, 2000). Nevertheless, this perspective does provide a holistic view regarding how children interact with the natural environment and can contribute to an understanding of the environment-human interrelationship (Clayton & Opotow, 2003).

2.3   School Visits to Botanic Gardens

For many teachers, the most important reason for undertaking botanic garden visits is that they offer the opportunity to address topics listed in the science and geography curricula (Jones, 2000). Consequently, often the learning activities organized either by schoolteachers or BGEs are focused on investigating issues, such as plant adaptation, and measuring different temperatures and humidity. However, during the visits, the children should not only obtain scientific and geographic knowledge but also be encouraged to develop their sense of social justice and moral responsibility as well as being taught to understand that their own choices and behaviour can affect local, national and global issues (QCA, 2000).

Botanic gardens are resources for environmental education in its broadest sense, as various elements of knowledge can be integrated within an excursion, for example: ecological literacy, environmental awareness and environmental sensitivity (Emmons, 1997; Hargreaves, 2005; Tal, 2004). Moreover, research has suggested that a school trip to a botanic garden should include 'not only knowledge and understanding of animals or plants groups, but also the process of science and general aspects, such as care for the environment and communication' (Tunnicliffe, 2001, p. 33). Further, Jones (2002) argued that 'a school visit to a botanical garden can encourage young people to think through their identity and place within society, both at the local and global level' (pp. 279–280). Moreover, a botanic garden can serve as the context for making these links and for implementing environmental, global and developmental education, a point illustrated by Jones (2003):

> Certainly the children that went to the garden were eager to think about where lots of products were from when they got back to school. They linked material products with plants and places and considered how these places were linked to both their schools and their homes. The other side of the world was seen as intimately linked with their everyday world and the botanical garden offered an exciting, interesting and colourful resource through which these experiences could be engaged with. (p. 29)

Most school visits to botanic gardens are usually one-day trips or last just a few hours and because of this limited period of time, the question arises as to how such a short experience can have an impact on children's learning, both cognitively and affectively. In this regard, in order to discover whether students' attitudes towards plants could be changed by visiting a botanic garden on a school trip, South (1999) asked primary students to draw a leaf at the beginning of a garden workshop and again after it. She found that 'there was an increase in the percentage of atypical leaves in the second set of drawings in all the classes' (p. 72) and thus she concluded that the botanic garden visiting experience had expanded students' observational views about plants. She also elicited that the impact of this on children in the age group 5–7 years old was less significant than that observed for the 7–9-year-olds. From this research, South (1999) suggested that if the botanic garden experience is to produce any significant impact on schoolchildren's environmental awareness, BGEs need to stimulate their interest by challenging their conceptual thinking.

Bowker and Jasper (2007) explored the conceptual learning of students who attended the BGEs' guided visits in the Eden Project in Cornwall and they adopted a personal meaning mapping (PMM) tool to measure how 'a specified learning experience uniquely affects each individual's meaning-making process' (Bowker & Jasper, 2007, p. 139). They asked 30 primary school students aged between 10 to 11 years old to describe a tropical rainforest, by writing and drawing on worksheets administered before and after the lesson. The instrument used for this (PMM) was based on the child-centred principle of focusing on the knowledge, feelings and perceptions that the children consider important. Furthermore, with regard to the PMM, Adams et al. (2003) have outlined its usefulness in measuring children's understanding along four semi-independent dimensions, those of extent, breadth, depth and mastery. In the work of Bowker and Jasper (2007), the analysis of the concept maps showed that children's understanding of tropical rainforests increased comprehensively after they had participated in the BGEs' guided lessons. In light of these results, they drew the conclusion that children can achieve learning even in the short amount of time available on a visit.

Some researchers have investigated the processes of how children learn about the environment during school trips to botanic gardens (Davies, Sanders, & Amos, 2015; Nyberg & Sanders, 2013). For example, Jones (2003) tracked more than 150 young people who visited the Birmingham Botanical Gardens and Glasshouses, with their school or family or as part of an out-of-school leisure group, and applied a range of qualitative research methods, such as participant observation, focus groups and text analysis. The findings of the study suggested that children learn better when teachers, BGEs, peers and chaperones are engaged in the activities. Furthermore, it was revealed that young people can use their previous knowledge to decide where to focus their attention so as to gain new insights. In general, they discovered that the experience of going to botanic gardens has a positive impact on young people's environmental understanding but of most significance is the part played by personal experience for developing a better understanding of the environment, for as one child who participated in the research reflected:

I think to learn you've got to have hands on experience. If you just learn from textbooks about the environment, say about how plants are grown, you don't actually look at them and you don't experience them. (quoted in Jones, 2003, p. 2)

Similarly, Stewart (2003) investigated the experiences of seven groups of primary and secondary children aged from 5 to 18, during their school excursion to the Royal Botanic Gardens in Sydney. Both pre- and post-visit interviews with the students ($n=50$) were conducted and a survey ($n=284$) about their visiting experience was also carried out. The author reported that school trips to botanic gardens usually involve two types of learning: learning for cognitive gains and for scheme-building, with the former referring to the measurable cognitive outcomes that students can achieve during tightly structured activities such as visits to specific displays to conduct specific tasks, whereas the latter is achieved when students can demonstrate long-term recall of plants, plant displays and specific locations at a botanic garden. Taken together, these two forms of learning can contribute to students' deeper understandings of plants, especially plant structure and biodiversity. In sum, Stewart (2003) proposed that practical activities, especially sensory experiences, form a key part of students' long-term recall of their botanic garden experiences.

Although botanic garden visiting experiences have a positive impact on s chil- dren's cognitive learning, some researchers have found that inappropriate teaching may lead to low levels. For example, Bowker (2004) studied a group of primary aged (7–11 years old) children who were led by a schoolteacher to the Eden Project in Cornwall, with the purpose being to elicit the most effective methods of utilizing a teacher-led school trip so as to enhance children's perceptions of plants and their understanding of people's relationships with them. Seventy-two participating children were interviewed within one month of the initial visit and the researcher discovered that they were affected by the sensory experience of being immersed in a garden with such a profusion of plants from around the world. However, although most of the children showed an interest in the plants that were relevant to their lives, it emerged that they were often unsure of the relationship between plants, people and resources. For example, just over 50 % of the children were able to articulate the link between plants and food, but only 33 % could make an unprompted link between plants and clothes. In light of this outcome, the researcher contended that to facilitate children's understanding of plants and the relationship that human soci- ety has with them, it is essential for the educator who is guiding the group during the visit to challenge students' ideas. This can be achieved by asking 'quality ques- tions that will focus children's attention on important aspects of plants such as plant adaptations to their climate or how people have used and cultivated certain plants' (Bowker, 2004, p. 240).

Similar results were reported by Tunnicliffe (2001), who explored the quality of primary school students' (aged 7–11) learning when they were looking at plant exhibits in the Royal Botanic Gardens at Kew, by collecting and analysing their conversations during the visit. The author found that their level of cognition was low, as they only 'talk spontaneously about the easily observed features of plants', but 'the functions of plants were hardly talked about, though a few conversations mentioned seed production and obtaining food' (Tunnicliffe, 2001, p. 32). In order to promote higher-order learning on the visits, the author proposed that teachers and BGEs' teaching should focus on a particular set of anatomical features and encourage students to construct their understandings through 'predicting, hypothe- sizing design observational protocols, gathering data and evaluating it' (Tunnicliffe, 2001, p. 33).

2.4 Teaching in Botanic Gardens: A Missing Pedagogy

When compared to the literature about school visits to botanic gardens, much more is known about visits to museum settings. Thus, after first reviewing the limited relevant literature about school groups in botanic gardens, we then examine the research that has been carried out in museums with the expectation of being able to highlight the aspects that are equally valid when applied to botanic gardens. Previous research on school visits to botanic gardens has mainly focused on students' learning experiences by highlighting the affective and cognitive gains (Bowker & Jasper, 2007), the diverse ways of interacting with plants for botanical learning (Sanders, 2007) and the learning process by analysing student-student interactions (Tunnicliffe, 2001). Moreover, Sanders (2007), when conducting a case study in the London Chelsea Physic Garden, reported that the predominant teaching approach used with visiting school groups was a mixture of traditional and enquiry-based teaching. Sanders criticized the pedagogical approach of botanic gardens towards school groups as being based on 'attitudes that focus on behaviour management and controlled didactic teaching and learning' (p. 1224).

Likewise, research on school groups in the museum setting has shown that some museum educators have failed to enrich students' learning by following a traditional knowledge-transmission model of teaching. For instance, Cox-Peterson, Marsh, Kisiel and Melber (2003) found that the museum educators in a US science museum used a lot of scientific jargon without providing students with analogies, information or explanations to relate the content knowledge to their lives outside the museum. It was also noted that the vast majority of the questions that these museum educators asked were closed and, once asked, lacked follow-up, elaboration or probing. Similar results were found in Tal and Morag's (2007) research on guided school visits in Israeli science museums. It was reported that the didactic way of teaching was commonly observed and when lecturing, the museum educators 'stayed at the centre, and rarely initiated discussion or listened to the students' questions and stories' (p. 763).

There is a growing trend to examine learning dialogues in research in both formal and informal education contexts; however, much of the research has focused its analysis on the particular linguistic forms or genre of discourse. Although there is an emerging body of research that focuses on the functions of family talk in museum settings (Palmquist & Crowley, 2007; Zimmerman, Reeve, & Bell, 2010), little is known about the functions of informal educators' talk during guided school visits. In order to address this gap, it is important for this study to examine the effectiveness of instructional discourse, which is determined by 'the quality of teacher-student interactions and the extent to which students are assigned challenging and serious epistemic roles requiring them to think, interpret, and generate new understandings' (Nystrand, 1997, p. 7).

References

Adams, M., Falk, J. H., & Dierking, L. D. (2003). Things change: Museums, learning, and research. In M. Xanthoudaki, L. Tickle, & V. Sekules (Eds.), *Researching visual arts in education in museums and galleries* (pp. 15–32). Dordrecht: Kluwer.

BGCI. (2008a). Definition of a botanic garden. Retrieved November 3, 2008, from http://www.bgci.org/botanic_gardens/1528/

BGCI. (2008b). The history of botanic gardens. Retrieved August 28, 2008, from http://www.bgci. org/botanic_gardens/history/

BGCI. (2008c). Mission statement. Retrieved November 27, 2008, from http://www.bgci.org/ global/mission/

Bowker, R. (2004). Children's perceptions of plants following their visit to the Eden Project. *Research in Science and Technological Education, 22*(2), 227–243.

Bowker, R., & Jasper, A. (2007). Don't forget your leech socks! children's learning at the Eden Project. *Research in Science and Technological Education, 25*(1), 135–150.

Boyd, W., & King, E. J. (1995). *The history of western education* (12th ed.). Lanham, Maryland: Barnes & Noble Books.

Brockway, L. H. (1979). Science and colonial expansion: The role of the British royal botanic gardens. *American Ethnologist, 6*(3), 449–465.

Clayton, S., & Opotow, S. (Eds.). (2003). *Identity and the natural environment.* Cambridge, MA: The MIT Press.

Comenius, J. A. (1660). *The school of infancy (edited with an introduction by E.M.Eller 1998.* Chapel Hill, CA: University of North California Press

Cox-Petersen, A. M., Marsh, D. D., Kisiel, J., & Melber, L. M. (2003). Investigation of guided school tours, student learning, and science reform recommendations at a museum of natural history. *Journal of Research in Science Teaching, 40*(2), 200–218.

Darling, J. (1994). *Child-centred education and its critics.* London: Paul Chapman.

Davies, P., Sanders, D., & Amos, R. (2015). Learning in cultivated gardens and other outdoor landscapes. In C. J. Boulter, M. Reiss, & D. Sanders (Eds.), *Darwin-inspired learning* (pp. 47–58). Rotterdam: Sense.

Dewey, J. (1916). *Democracy and education: An introduction to the philosophy of education.* New York: Macmillan.

Dewey, J. (1938). *Experience and education.* New York: Macmillan.

Emmons, K. M. (1997). Perceptions of the environment while exploring the outdoors: A case study in Belize. *Environmental Education Research, 3*(3), 327–344.

Falk, J. H., & Dierking, L. D. (2000). *Learning from museums: Visitor experiences and the making of meaning.* Walnut Creek, CA: AltaMira Press.

Hargreaves, L. J. (2005). *Attributes of meaningful field trip experiences* (Unpublished Master's thesis). Simon Fraser University, Vancouver, Canada.

International Union for Conservation of Nature and Natural Resources. (1980). *World conservation strategy: Living resource conservation for sustainable development.* Gland, Switzerland: International Union for Conservation of Nature and Natural Resources (IUCN).

Jones, V. (2000). *More than just plants: Engaging with the politics of identity at botanical gardens.* Paper presented at the Making Sense of Teaching and Learning Through Environmental Education Research, Chelsea Physic Garden, London.

Jones, V. (2002). Identity and the environment. *The Curriculum Journal, 13*(3), 279–288.

Jones, V. (2003). *Young people and the circulation of environmental knowledge* (Unpublished doctoral dissertation). University of Wales, Aberystwyth, UK.

Montessori, M. (1912). *The Montessori method* (A. E. George, Trans.). New York: Frederick A. Stokes Company.

Nyberg, E., & Sanders, D. (2013). Drawing attention to the 'green side of life'. *Journal of Biological Education, 48*(3), 142–153.

Nystrand, M. (1997). *Opening dialogue: Understanding the dynamics of language and learning in the English classroom.* New York: Teachers College Press.

Palmquist, S., & Crowley, K. (2007). From teachers to testers: How parents talk to novice and expert children in a natural history museum. *Science Education, 91*(5), 783–804.

QCA. (2000). National curriculum: Citizenship key stage 2. Retrieved October 28, 2008, from http://curriculum.qca.org.uk/key-stages-1-and-2/subjects/citizenship/keystage2/index. aspx?return=/key-stages-1-and-2/subjects/citizenship/index.aspx

Rae, D. (1996). *Botanic gardens and their live plant collections: Present and future roles* (Unpublished doctoral dissertation). The University of Edinburgh, Edinburgh, UK.

Rowe, S., & Humphries, S. (2004). The outdoor classroom. In M. Braund & M. Reiss (Eds.), *Learning science outside the classroom* (pp. 19–34). London\New York: Routledge Falmer.

Sanders, D. (2004). *Botanic gardens: 'Walled, stranded arks' or environments for learning?* (Unpublished doctoral dissertation). University of Sussex, Brighton, UK.

Sanders, D. (2007). Making public the private life of plants: The contribution of informal learning environments. *International Journal of Science Education, 29*(10), 1209–1228.

Sanders, D. (2008). (Personal communication).

Sealy, M. R. (2001). *A garden for children at Family Road Care Center* (Unpublished Master's thesis). Louisiana State University and Agricultural Mechanical College, Baton Rouge, LA.

Smith, M. K. (2008). Fredrich Froebel. Retrieved May 10, 2008, from http://www.infed.org/thinkers/et-froeb.htm

South, M. (1999). Can a botanic garden education visit increase children's environmental awareness? In L. A. Sutherland, T. K. Abraham, & J. Thomas (Eds.), *The power for change: Botanic gardens as centres of excellence in education for sustainability. Proceedings of the 4th International Congress on Education in Botanic Gardens* (pp. 68–76). Richmond, Surrey: Botanic Gardens Conservation International.

Stewart, K. M. (2003). *Learning in a botanic garden : The excursion experiences of school students and their teachers* (Unpublished doctoral dissertation). University of Sydney, Sydney, Australia.

Subramaniam, A. (2002). Garden-based learning in basic education: A historical review. *Monograph.* http://fourhcyd.ucdavis.edu

Tal, T. (2004). Using a field trip to a wetland as a guide for conceptual understanding in environmental education: A case study of a pre-service teacher's research. *Chemistry Education: Research and Practice, 5*(2), 127–142.

Tal, T., & Morag, O. (2007). School visits to natural history museums: Teaching or enriching? *Journal of Research in Science Teaching, 44*(5), 747–769.

Tunnicliffe, S. D. (2001). Talking about plants: Comments of primary school groups looking at plant exhibits in a botanical garden. *Journal of Biological Education, 36*(1), 27–34.

World Commission on Environment and Development [WCED]. (1987). *Our common future: Report of the World Commission on Environment and Development (WCED).* Oxford: Oxford University Press.

Wyse-Jackson, P. S., & Sutherland, L. A. (2000). *International agenda for botanic gardens in conservation.* Surrey, UK: Botanic Gardens Conservation International (BGCI).

Zimmerman, H. T., Reeve, S., & Bell, P. (2010). Family sense-making practices in science center conversations. *Science Education, 94*(3), 478–505.

Chapter 3
A Sociocultural Perspective of Teaching and Learning in Formal and Informal Science Settings

Drawing on a sociocultural perspective of teaching and learning, this chapter underpins the importance of talk in structuring thinking and shaping higher mental processes. More specifically, through dialogic teaching, educators can create interactive opportunities for students to negotiate meanings and scaffold their understandings. The relevant literature regarding research pertaining to both formal and informal contexts is reviewed to identify the different functions of educators' talk in supporting students' learning. A general conclusion can be drawn that effective interactions between educators and children, especially through questioning, have a significant impact on the level of learning.

3.1 The Role of Talk in Teaching and Learning

A sociocultural perspective on learning is adopted that originates from Vygotsky's theoretical work on child development in which he advanced the concept that learning and development involve a passage from social contexts to individual understanding, that is, knowledge is first encountered in interactions between people and then internalized into the learner's repertoire of understanding (Kozulin, 2003; Wertsch, 1991). From this perspective, language is characterized as being the key instrument for cultural transmission and psychological mediation. As a cultural tool, language enables social knowledge to be transmitted to future generations, whilst when it serves its psychological functions, it allows social participants to organize their thoughts, reason, plan and review their actions.

According to Vygotsky (1987), these dual functions of language are integrated. With respect to this integration, Mercer (2000) considered the process through language for human individuals and their societies as one in which 'children hear people in their communities using language to describe experience and get things done, they pick up these cultural ways with words and eventually make them their own psychological tools' (p. 10). That is, through talking with other people, children make sense of the world and gain the communication skills required to become active members of their communities, a process succinctly summed up by Rogoff (1990) in the following statement: 'interaction with other people assists children in their development by guiding their participation in relevant activities, helping them adapt their understandings to new situations, structuring their problem-solving attempts and assisting them in assuming responsibility for managing problem solving' (p. 191). Therefore, talking with adults and peers plays a significant role in children's intellectual and social development.

The advocates of the sociocultural perspective of learning have argued that knowledge is not constructed through interaction, but rather, in interaction (Wertsch, 1991, 1998). Turning to an educational context, this stance towards interaction, as regarded under a sociocultural lens, suggests that classroom discourse is not effective unless students play an active part in their learning through exploratory forms of talk (Barnes, 2008). Moreover, when students are given greater control by being able to initiate ideas and come up with responses, they maximize the potential for developing shared understandings (Edwards & Westgate, 1994; Nystrand, 1997). In keeping with this contention, a significant body of research has established that teacher-student and student-student interactions during classroom activities have great potential regarding learner engagement in constructing shared knowledge, providing that such interactions are learning provoking and dialogic in nature (e.g. Alexander, 2006; Mercer & Littleton, 2007; Nystrand, Wu, Gamoran, Zeiser, & Long, 2003). Further, Mercer and Littleton (2007) argue that the success of education 'may be explained by the quality of educational dialogue, rather than simply by considering the capacity of individual students or the skill of their teachers' (p. 4). Thus, the quality of classroom interaction provided by teachers should be an important criterion in evaluating learning, especially in their formative assessment, as the feedback this involves can enhance their understanding of the subject matter (Bell & Cowie, 2001; Black & Wiliam, 2001). Atkin and Black (2003) emphasis the role of formative assessment in supporting children to move forward in their learning and noted:

> What really counts in education is what happens when teachers and pupils meet. The wisdom of any decision about education is best judged on the basis of whether or not it raises the quality of these interactions. (p. xi)

The interest regarding dialogic interactions in teaching and learning has generated a new stream of educational research, the focus of which requires scholars to understand how classroom discourse develops and its essential role in engaging students as well as supporting their learning.

3.1.1 Dialogic Interaction in Teaching and Learning

Although nowadays students are usually no longer perceived as being mere passive recipients but, rather, active participants in constructing knowledge, some researchers have observed that teacher-centred, recitation-based and rote teaching are still prevalent in American and English classrooms (Nystrand, 1997; Smith, Hardman, Wall, & Mroz, 2004). In such studies, the predominant discourse in these class-rooms was found to involve a triadic pattern of interactions, consisting of an initia-tion by the teacher, followed by a response from a student, with subsequent evaluation or feedback to the student's response from the teacher (I-R-E or I-R-F). Some research on the triadic pattern of interactions has shown that speech-like instructions given by the teacher will limit the opportunity for students to think, explain and generate new understandings, whereas these objectives can be achieved through dialogic teaching (Alexander, 2006), as discussed next.

Dialogic teaching focuses on the interactions between the teacher and students and can provide structured, extended process leading to new insights and deep knowledge and understanding and, ultimately, better practice, and consequently, educators should be encouraged to support their learners to engage in building a continuous rapport (Alexander, 2008). With regard to the language of dialogue, Cazden (2001) has specifically identified its function as comprising the following: (a) the communication of propositional information, (b) the establishment and maintenance of social relationships and (c) the expression of the speaker's identity and attitudes. Furthermore, she went on to explain that the propositional, social and expressive functions of language align with how the cognitive and the social dimen-sions of being are integrated and how an individual's identity and attitudes come to be developed. Thus, regarding school education and education in out-of-classroom contexts, educators have a significant responsibility for stimulating and supporting higher-order thinking, through engaging their students in appropriate educational dialogues (DeWitt & Osborne, 2007; Mercer, 2008).

Turning to the empirical evidence on dialogic practice, during an investigation of the guided participation taking place in some Mexican primary classrooms, Rojas-Drummond and colleagues (Rojas-Drummond, 2000; Rojas-Drummond, Mercer, & Dabrowski, 2001) found that the students' competency and independency in prob-lem solving and reasoning were influenced by the ways in which teachers interacted with them. More specifically, it emerged that compared with the teachers whose students achieved lower scores, those in similar schools, whose students got better results, tended to provide a social-constructivist approach to teaching and learning. That is, in the more efficient teachers' classrooms, questions were posed not just to elicit students' knowledge but also to guide the development of reasoning for under-standing. Moreover, rather than conveying the subject content directly, these teach-ers encouraged students to make explicit their own thoughts by elaborating upon their process of arriving at solutions. In addition, students were encouraged to take a more active, vocal role in the various classroom activities, through exchanging ideas with the whole class. As Rojas-Drummond and Mercer (2003) have argued, by using the appropriate interactional strategies, teachers can allow students 'to become more able in managing individual and joint reasoning and learning activi-ties in the classroom' (p. 99).

According to Ash and Wells (2006), dialogic inquiry should be the fundamental practice for education in both formal and informal contexts. They argue that in both contexts, social interactions can promote collaborative knowledge building, mediated by artefacts and dialogue, when answers to questions are not determined in advance and when expertise is distributed. Research on family groups in informal science settings has revealed that adults and children can genuinely explore the environment and reflect on each other's ideas when the former function as more experienced and responsible mentors rather than as authoritative teachers (Tal, 2012). Classroom research has shown that the way in which a teacher responds to students' contributions directs the development of any discussion (Hardman, 2008; Nystrand et al., 2003; Wells & Arauz, 2006). Therefore, it is important that teachers are aware of the pedagogical functions of talk and how it can best be used to facilitate students' learning (Mercer & Howe, 2012). For advancing the collaborative discussion, a teacher should not act as the sole source and arbiter of knowledge but seek to elicit different ideas whilst at the same time supporting students to express their contributions (Ash & Wells, 2006). Thus, teachers need to balance strategically their authoritative talk with dialogue so as to keep students actively engaged in building knowledge and expertise that can become widely distributed among the members of the class (Mercer & Howe, 2012; Scott, Mortimer, & Aguiar, 2006).

3.1.2 Pedagogical Functions of Educators' Talk

Although research regarding classroom communication started in the 1960s, few scholars have succeeded in drawing out the connections between classroom discourse and the processes of teaching and learning. One pioneering study, carried out by Sinclair and Coulthard (1975), was influential in the field because it provided a groundbreaking systematic exploration of the interactions between teachers and students through a rigorous linguistic analysis of their patterns. More specifically, they developed an analytical scheme that categorized classroom talk under the headings of 'lesson', 'transaction', 'exchange', 'move' and 'act'. According to these authors, an analytical scheme, such as theirs, can be relied upon to capture all the talk occurring in lessons for the purposes of further research and analysis (Sinclair & Coulthard, 1975). However, although this particular analytical framework offers a means to analyse classroom discourse, its main limitation is that it only focuses on the linguistic functions of teacher-student interactions and, hence, fails to incorporate the pedagogic dimensions of the interactions. A systematic classroom observation schedule that offered opportunities for a richer analysis was advanced by Eggleston, Galton and Jones (1976). Their schedule explored the intellectual transactions taking place during the processes of teaching and learning and was developed by these researchers in the context of observing science lessons (see Table 3.1).

In their seminal book *CommonKnowledge*, Edwards and Mercer (1987) linked classroom discourse to its educational functions and reported that teachers used different approaches to fulfil their educational purposes. These approaches include paraphrasing students' contributions, offering reconstructive recaps and direct lecturing. Moreover, Mercer (1995), from his studies of teacher talk in classroom

Table 3.1   Science teaching observation schedule used by Eggleston et al. (1976)

Type of teacher talk	Characteristics of teacher talk
Teacher questioning for	Recalling facts or principles
	Applying facts and principles to problem solving
	Making hypotheses or speculation
	Designing experiments
	Direct observation
	Interpretation of observed or recorded data
	Making inferences from observations or data
Teacher statement about	Fact and principle
	Problems
	Hypothesis or speculation
	Experimental procedure
Teacher response to students for	Acquiring or confirming facts or principles
	Identifying or solving problems
	Making inferences, formulation or testing hypotheses
	Seeking guidance on experimental procedure

setting, suggested that there are three specific techniques available to teachers for guiding children's learning: eliciting knowledge from students, responding to what students say and describing shared classroom experience. The first includes two approaches, direct and cued, both of which comprise a process through which the students actively participate in the creation of shared knowledge, rather than merely sitting and listening to the teacher. That is, the purpose of elicitation is to assist students to move from within their 'zone of proximal development' (Vygotsky, 1978) towards the more challenging end move, with the help of the teachers' prompts, cues and questions. Regarding the second matter, that of responding to what students say, this refers to the teacher not only giving feedback to what students say but the teacher incorporating their contributions into the flow of classroom discourse so as to construct more generalized meanings. More specifically, teachers can use a range of strategies to respond including: 'confirmation', 'rejection', 'repetition', 'elaboration' and 'reformulation' (Mercer, 1995, p. 34). By repeating what the students say, the teacher can draw the whole class's attention to an answer and/or emphasize the educational significance of that remark, whereas by paraphrasing or reformulating students' utterances, the class can be given modified, clarified and accurate versions of subject matter. Elaboration is another way through which the teacher can expand or explain a statement from a single student to the whole class. The third technique involving the shared classroom experience requires the teacher to base the future talk, activity and learning on that shared experience. To achieve this, Mercer (1995) has advised the use of 'we' statements, which he contended effectively develop students' awareness of having common past experiences as they gain shared knowledge and collective understanding. In sum, these techniques provide teachers with a linguistic tool kit to frame classroom discourse, thereby promoting shared understanding of the thematic content of the lessons. That is, as Mercer (1995) argued, such techniques enable the teacher:

> to help learners appreciate the relevance of their existing knowledge, to help them realize what they know, to help them see continuities in their experiences and to introduce them to new knowledge in ways that allow them to make sense of it in terms of what they already know. (p. 38)

In addition to the suggested techniques of Edwards and Mercer (1987) and Mercer (1995) for analysing teacher talk, O'Connor and Michaels (1996) carried out a study which focused on how a teacher elaborates or reformulates what students say in a classroom discourse activity. They used the term 're-voice' to express how a teacher reutters a student's contribution through the use of repetition, expansion and rephrasing. In re-voicing a student's contribution, the teacher may add or delete material and use different words or phrases, in order to clarify, highlight or reframe aspects of the student's utterance in relation to the current or desired academic content of knowledge. By using these strategies, the teacher is able to 'place one pupil in a relation to other pupils as holders of positions' (O'Connor & Michaels, 1996, p. 77), that is, they can provide the student with a stance with respect to the topic under discussion, whilst continuing to engage the rest of the class in a relationship consisting of a potentially extended discussion. Thus, re-voicing can be considered as a strategy that promotes academic debate by showing how their ideas relate to those of others. Moreover, rather than simply promoting argumentation and debate, functions of re-voicing have been identified as: (a) allowing the teacher effectively to credit a student for his or her contribution, whilst still clarifying or reframing the contribution in terms most useful for group consumption; (b) socializing students into particular intellectual and speaking practices by placing them in the roles entailed by the speech activity of group discussion; and (c) bringing students to see themselves and each other as legitimate participants in the activity of making, analysing and evaluating claims, hypotheses and predictions. In re-voicing, students are animated as theorizers, predictors or hypothesizers, thereby crediting them with being the major driving force of classroom discourse. However, during such circumstances it is still the teacher who controls the talk and creates the dramatic landscape where they are embedded 'in the ongoing activity and in terms of the actual propositional content under discussion' (O'Connor & Michaels, 1996, p. 78).

The pattern of the teacher talk has a great influence on students' conceptual understanding of scientific concepts and procedures. In this regard, Lemke (1990) argues that 'learning science means learning to talk science', in other words, students are 'learning to communicate in the language of science and act as a member of the community of people who do so' (p. 1). From this, it follows that the aim and purpose of science education is to help students learn to talk science through classroom discourse, and in order for this to happen, he proposed a set of strategies that teachers can adopt when in dialogue with students. These strategies are divided into two, according to the nature of the teacher talk involved, namely, dialogue and monologue strategies. Furthermore, Lemke (1990) goes on to address the function of teacher questions in teaching, for which he proposed the construction of a set of linked semantic relations, 'organized according to particular rhetorical and genre patterns and sometimes even realized by the same lexicogrammatical means, from one text to the next, from one occasion of discourse to another' (p. 36). To build up a network of semantic relations in the dialogue, the teacher can select and modify student answers for further discussion. In addition, the teacher can recontextualize the students' contributions to the classroom discourse through elaboration and, in so doing, place them in a different thematic context. With regard to monologue strategies, these can include: logical exposition, narrative, selective summary and foregrounding and backgrounding. More specifically, through logical exposition, the teacher can make logical connections between various thematic items and semantic relations with a narrative which can serve to 'link the thematic processes and items of one narrative event or episode to those of the next, creating a complete exposition of the thematic pattern' (Lemke, 1990, p. 109). Selective summary refers to the teacher summarizing the prior discourse, which can include selected thematic elements and relations, and finally, by foregrounding and backgrounding, the teacher can repeat or summarize prior discourse to emphasize the degree of importance of a certain theme.

It is apparent that there are some overlaps between Lemke (1990) and Mercer's (1995) perspectives regarding the nature of teacher talk, the latter's having been presented earlier in this chapter. First of all, both of them centralized the role of the teacher asking questions in student knowledge building and both recommended using cued elicitation. Furthermore, they were in agreement that the strategies identified in teacher talk, such as selective summary and foregrounding, can contribute to the creation of shared experience.

Drawing on the analyses that focused on pedagogical interventions by Mercer and Edwards (Edwards & Mercer, 1987; Mercer, 1995) and Lemke (1990), Scott (1998) has identified five strands of pedagogical intervention in relation to how teacher talk supported students' meaning making in secondary science classroom settings. Each strand of intervention could be characterized by the narratives or teaching performance, adopted by a teacher so as to direct and sustain students in their sense making regarding scientific understandings. The detailed pedagogical interventions identified in Scott's framework are summarized below in Table 3.2.

This framework of pedagogical intervention shows that teacher talk can serve the purposes of several different strands of teaching narrative simultaneously. For example, a teacher's action to ask a student to clarify an idea can fall into the two categories of 'checking student understanding' and also that of 'promoting shared meaning'. Hence, it can be concluded that the teacher talk perspectives described above are not isolated, serving a single purpose, but can be used as techniques to support and promote students' learning.

In informal settings, such as museums, the educators' talk also plays an important role in shaping children's visiting experiences. Based on the description of teacher talk in previous research (Edwards & Mercer, 1987; Lemke, 1990; Mercer, 1995; O'Connor & Michaels, 1996; Scott, 1998; Sinclair & Coulthard, 1975) and observation of museum explainers' teaching practices, King (2009) developed a

Table 3.2 A framework of the teaching narrative in science education (Adapted from Scott, 1998)

Strand	Teacher performance/teaching narrative
Developing the conceptual line	Shaping ideas: guiding students through the steps of an explanation by means of a series of key questions; paraphrasing students' ideas; differentiating between ideas
	Selecting ideas: selecting a student idea or part of a student idea; retrospectively eliciting a student idea; overlooking a student idea
	Marking key ideas: repeating an idea; asking a student to repeat an idea; enact a confirmatory exchange with a student; validating a student idea; posing a rhetorical question; using a particular intonation of voice
Developing the epistemological line	Introducing students to aspects of the nature of scientific knowledge (e.g. the generalizability of scientific explanations)
Promoting shared meaning	Presenting ideas to the whole class
	Sharing individual student ideas with the whole class
	Sharing group findings with the whole class
	Jointly rehearsing an idea with a student in front of the whole class
	Providing a spoken commentary to make explicit the thinking behind a specific activity that they are engaged in
Checking student understanding	Asking for clarification of a student idea
	Checking individual student understanding of particular ideas
	Checking the consensus in the class about certain ideas
Maintaining the narrative	Stating aims/purposes for the next part of the narrative
	Looking ahead to anticipate possible outcomes
	Reviewing progress of the narrative
	Refocusing the discussion

coding scheme to identify their talk, which includes 12 kinds of talk, including: cue, withhold answer, select, non-select, repeat, re-voice, distinguish, align, model, create shared experience, seek to inspire and explain. This provides a thorough framework for analysing the role of educator talk in informal settings and is adapted to analyse the BGEs' talk in this study, with the modified analytical framework being presented in Chap. 4.

3.1.3   Educator Questioning

Questioning is one of the most widely used techniques employed by educators, regardless of their views on learning or lesson planning. In reality, there are many practical reasons for teachers to ask questions in a classroom setting, for example: to keep students active and attentive, to check knowledge or understanding, to stimulate curiosity and interest, or to diagnose problems in learning. The initial research investigating teacher questioning conducted in the 1980s tended to consider the learning outcomes of students as a function of their questioning practices

(Dillon, 1982; Mehan, 1979; Redfield & Rousseau, 1981), and since then, this somewhat process-product approach has been criticized for being too simplistic, in particular, because of its tendency to ignore the contexts in which the questions were posed (Edwards & Mercer, 1987). Consequently, more recently, the focus of research endeavours has turned towards the investigation of the understanding of contextual issues, such as the context of questions and their content (Carlsen, 1991).

With regards to the usage of questions in science classroom settings, van Zee and Minstrell (1997) have identified and prompted the use of the 'reflective toss', which comprises three separate utterances: a student statement, teacher question and an additional student statement. The metaphor 'toss' refers to the teacher using questions to 'catch the meaning' of the student's prior utterance and by so doing throwing the responsibility for thinking back to the students (van Zee & Minstrell, 1997, p. 229). That is, rather than evaluating student statements, the teacher asks questions to help them clarify their meanings, consider different viewpoints and monitor their discussion and thinking. This kind of questioning approach not only seeks to elicit what students think but also furnishes the classroom discourse with a dialogic element that promotes the construction of 'common knowledge' between the teacher and the students (Edwards & Mercer, 1987).

Teacher questioning not only invites students to participate in classroom discussion but also promotes their conceptual thinking. In this regard, Chin (2007) has identified four types of questions used by science teachers to stimulate students' productive thinking, namely: Socratic questioning, semantic tapestry, verbal jigsaw and framing as set out in Table 3.3. Socratic questioning refers to those questions that are used to probe, extend and elaborate students' thinking, rather than directly offer up information, whereas semantic tapestry includes those questions that the teacher deploys to develop students' conceptual and relational understandings of abstract concepts. Questions under the heading verbal jigsaw are those posed with the purpose of guiding students in developing their understanding and use of the various scientific terminologies, key words or phrases. Lastly, framing refers to the questioning technique in which the teacher frames a problem, issue or topic and thus guides the discussion that flows from it.

Some scholars investigated how teachers respond to the feedback from their students after posing a question and proposed that appropriate teacher feedback, which aims to encourage and extend student contributions, can promote higher levels of interaction and cognitive engagement (Hardman, 2008; Nassaji & Wells, 2000; Nystrand et al., 2003). Regarding this, Chin (2006) conducted a qualitative investigation into how teacher feedback to students' responses can make classroom discourse more thought provoking and can stimulate more elaborate and productive student responses. She developed a framework containing four teacher feedback approaches, based on the accuracy of students' responses to the teachers' questions (see Table 3.4).

Using the framework above, Chin (2006) investigated the teachers' methods for responding to their students' answers and found that the 'affirmation-direct instruction' and 'explicit correction-direct instruction' feedback approaches failed to encourage student input beyond the initial solicited answers, whereas the 'focusing

Table 3.3 Teacher questioning approaches to stimulate productive thinking (Adapted from Chin, 2007)

Productive thinking stimulating questions	Questioning approach	The features of teacher questions
Socratic questioning	Pumping	Encouraging students to further articulate their thoughts and ideas through explicit requests and giving positive feedback
	Reflective toss	Posing a question in response to a previous utterance made by a student in order to throw the responsibility of thinking back to the students
	Constructive challenge	Posing a question to challenge students' thinking rather than giving them direct corrective feedback
Semantic tapestry	Multipronged questioning	Posing questions from different aspects of a problem to stimulate students to think deeply
	Stimulating multimodal thinking	Pose questions to articulate students' ideas in different forms (verbal, diagrams, visual images, symbols), which encourages students to think in a variety of modes
	Focusing and zooming	Posing questions that encourage students to think at both the macro- and micro-level or to zoom in and out, alternating between a big broad question to a more specifically focused question
Verbal jigsaw	Association of keywords and phrases	Posing questions to identify and articulate the keywords or phrases associated with the topic, which helps students to master the salient concepts and important scientific vocabulary
	Verbal cloze	Pausing in mid-sentence to allow students to fill in the blanks to complete it
Framing	Question-based prelude	Posing questions as a preface to subsequent presentation which may focus student thinking
	Question-based outline	Presenting a big, broad question and subordinate or related questions, which helps students to see the links between the big question and the subordinate ones
	Question-based summary	Giving an overall summary question to consolidate the key points

and zooming' and 'constructive challenge' strategies were effective in getting students to 'formulate hypotheses, predict outcomes, brainstorm ideas, generate explanations, make inferences and conclusions, as well as to self-evaluate and reflect on their own thinking' (p. 1336). The scholar concluded that if the teachers' feedback could be more facilitative rather than evaluative, the classroom discourse would become more thought productive by stimulating more elaborated student responses.

Table 3.4 Teachers' feedback to students' response (Adapted from Chin, 2006)

Nature of students' response	Types of feedback	Description
Correct	Affirmation-direct instruction	Affirm and reinforce response followed by further exposition and direct instruction
Mixture of correct and incorrect	Focusing and zooming	Accept response followed by a series of related questions that build on previous ones to probe or extend conceptual thinking
Incorrect	Explicit correction-direct instruction	Explicit correction followed by further expounding of the normative ideas
	Constructive challenge	Evaluative or neutral comment followed by reformulation of the question or challenge via another question

Compared to the abundant research regarding teacher questioning conducted in school classrooms, little is known about how educators use questions to support students' learning in informal settings. Tal and Morag (2007) explored the questions that museum educators asked students on school trips to a natural history museum, finding that they tended to ask rhetorical questions which they answered themselves, rather than waiting for a response from the children. Moreover, most of their questions were factual and demanded knowledge recall, with only a small proportion challenging students' thinking. In a comparative study of teacher-student talk that took place in a science classroom and on a museum trip, DeWitt and Hohenstein (2010) elicited that the teachers' closed-ended, task-related or procedural questions tended to predominate in both settings. However, the authors noticed that the teachers asked more 'request help' and 'invite participation' questions in the museum than in the classroom setting which the authors suggested involved a 'less dominant teacher role and a more balanced and interactive relationship, in which the students and teacher support each other' (DeWitt & Hohenstein, 2010, p. 466).

3.1.4   Student Questioning

Asking questions is an important part of our daily life, as 'to question is to ponder, to seek answers to a puzzle or a problem, to encounter a perplexity that requires resolution, to call something into question is to express doubt about it and to challenge its authenticity' (Pedrosa de Jesus, Teixeira-Dias, & Watts, 2003, p. 1017). In the educational context, King (1990) described students asking questions as a process through which learners:

> externalize their thoughts, making their ideas explicit and accessible both to themselves and to others in their group. Continued questioning and responding could guide group members to resolve these socio-cognitive conflicts by providing them opportunities to fill in the gaps in their knowledge structures, correct understandings, discover and resolve discrepancies in information and reconcile conflicting views. (p. 666)

Cuccio-Schirripa and Steine (2000) contend that 'questioning is one of the thinking processing skills which is structurally embedded in the thinking operation of critical thinking, creative thinking and problem solving' (p. 210). In spite of the fact that student-generated questions play an important role in teaching and learning, research based in science classrooms has shown that students seldom ask any questions, let alone on-task, high-quality questions (Dillon, 1988; Good, Slavings, Harel, & Emerson, 1987; Graesser & Person, 1994).

Questions formulated by students have the following functions: (a) to facilitate knowledge construction and enhance understanding (Chin & Brown, 2000; Harper, Etkina, & Lin, 2003; Scardamalia & Bereiter, 1992); (b) to engender discussion and debate and thus foster shared understanding (Aguiar, Mortimer, & Scott, 2010; Chin, Brown, & Bruce, 2002; Osborne, 2005); (c) to enhance students' interest and direct them to become autonomous learners (Chin & Kayalvizhi, 2005; Marbach-Ad & Sokolove, 2000; Ng-Cheong & Chin, 2009; Pedrosa de Jesus, Almeida, & Watts, 2004); (d) to enable teachers to diagnose students' understanding (Harper et al., 2003; Maskill & Pedrosa de Jesus, 1997; Watts, Gould, & Alsop, 1997); and (e) to assess their higher-order thinking (Dori & Herscovitz, 1999). Educators have the responsibility to encourage students to generate meaningful and productive questions in order to achieve these outcomes.

The questions that are asked by students have been analysed and subsequently classified by various scholars according to their perspectives on learning. For instance, student questions have been grouped according to the level of cognitive process required to answer them. The most famous and broadly acknowledged classification of human cognitive processes is Bloom's taxonomy, which although designed for teaching, has been suggested as a cognitive hierarchy model that could be used for the categorization of student-generated questions (Chin & Osborne, 2008; King, 1990). In his taxonomy, learning objectives were distinguished in an ordered hierarchy of level of thought, moving upwards from knowledge, comprehension, application, analysis and synthesis to evaluation. Following this and nearly a half century later, Anderson and Krathwohl (2001) modified this taxonomy of the cognitive domains of learning and formulated an ordered hierarchy comprising the dimensions of, from the bottom, remembering, understanding, applying, analysing, evaluating and creating.

Watts et al. (1997) classified three types of student-generated questions during the process of conceptual change being undertaken by students. They termed those questions produced when students are attempting to confirm explanations and consolidate understanding of their ideas as consolidation questions, whilst when students have reached a sense of conviction in their understanding, exploration questions are generated that help them to expand knowledge and test constructs. Finally, elaboration questions are asked by students when they attempt to reconcile different understandings, force issues and track in and around ideas and their consequences. This account regarding student questions implies that teachers should be aware of the nature of student-generated questions, as these may indicate the progression of students' conceptual understandings. Further, encouraging students to ask exploration and elaboration questions is a challenging task, because as Watts et al. (1997) noted, there is often a considerable barrier for students to ask questions aloud in class for fear of being considered as being 'stupid' or a 'boffin' by their peers. Since posing questions in class can generate feelings of exposure and vulnerability, teachers need to cultivate a learning environment where students can feel comfortable about asking questions.

Whether a child asks confirmation or transformation questions has been identified as a bipolar construct depending on 'the nature of the situation' on the one hand and their 'preferred style of working and the requirements of the task in hand' on the other (Pedrosa de Jesus et al., 2003, p. 1028). Using this bipolar construct, a typology of confirmation and transformation questions was developed, with the former referring to questions that seek to clarify information and ask for exemplification or definition, whereas the latter involves restructuring or reorganization of students' understanding through processes such as: exploring argumentative steps, identifying omissions, examining structures in thinking and challenging accepted reasoning. Pedrosa de Jesus et al. (2003) concluded that confirmation and transformation questions complement each other, and combined usage of the two forms indicates that a student is proficient in high-quality questioning.

The source of students' questions usually originates in there being a gap or discrepancy in their knowledge or alternatively because the questioner has the desire to extend his/her knowledge in some precise direction. More specifically, drawing on the potential for there being different underlying sources, Scardamalia and Bereiter (1992) distinguished two types of questions that students asked during science lessons. The first type that they identified is text-based questions, where students are instructed to generate questions in response to certain cues as part of their study of a text, whilst the other from knowledge-based questions occur spontaneously and are generated from a deeper interest of students in wanting to make sense of the world.

Building on Scardamalia and Bereiter's (1992) study, Chin et al. (2002) divided student-formulated questions during learning about science into two types, basic information and wonderment. The former are further separated into factual and procedural questions, with factual questions being those that are usually closed in nature and used to recall information, whereas procedural ones are produced when students seek clarification about the procedures that are required in order to complete a given task. Turning to wonderment questions, as Chin et al. (2002) explained that these 'were pitched at a conceptually higher level, required an application or extension of taught ideas and focused on predictions, explanations and causes instead of facts, or on resolving discrepancies and gaps in knowledge' (p. 531). Furthermore, they come to students' minds when learners attempt to 'relate new and existing knowledge, or build internal associations among different aspects of the new knowledge in their efforts to understand' (Chin et al., 2002, p. 531). Set out in the table below is Chin et al.'s typology of student-generated questioning (Table 3.5).

Moreover, from their observations of students who were engaged in hands-on laboratory activities in small groups, Chin et al. (2002) found that basic information questions, especially procedural questions, dominated in their conversations. Further, in spite of there being only a small proportion of wonderment questions, their data demonstrated that such questions did in fact stimulate students to generate explanations and solutions to problems. More specifically, the researchers observed

Table 3.5 Types of student-generated questions in learning science (After Chin et al., 2002)

Question	Sub-question	Characteristics
Basic information question	Factual question	Closed questions, to recall information
	Procedural question	To seek clarification of how to complete a task
Wonderment question	Comprehension questions	To seek an explanation of something that is not understood
	Prediction question	To predict a phenomenon according to observations, involving some speculation or hypothesis verification
	Anomaly detection question	To address scepticism or some discrepant information which causes cognitive conflict
	Application question	To find out what use is the information that the student himself/herself is dealing with
	Planning/strategy question	To wonder how best to proceed next when no prior procedure has been given

that through asking wonderment questions, the students were able to 'initiate a process of hypothesizing, predicting, thought experimenting and explaining, thereby leading to a cascade of generative activity' (Chin et al., 2002, p. 540). That is, wonderment questions asked by the students reveal a deeper approach to learning science that can trigger their engagement in productive discussions and thereby improve their conceptual understanding of science.

It is arguable that informal science settings may allow students to have sufficient time to develop their thoughts and ask questions. In this regard, following an investigation of science museum educators' guided school visits, Cox-Petersen, Marsh, Kisiel and Melber (2003) reported that student-generated questions were rarely observed, and the reason for this was that the museum educators did not allow sufficient time for students to formulate their own questions. It was found that approximately 20 % of the schoolteachers reported that they would have liked the tours to have given more time for student questions and enquiries, which underlines the point that when organizing learning-oriented school trips to informal settings, as Griffin and Symington (1997) have contended, staff need to 'use a learner-centred approach in which the students are finding questions to their own answers, rather than their teachers' or the museum's questions' (p. 777).

Moreover, it is believed that informal settings may provide students with more freedom to seek information from adults by asking questions. Tunnicliffe, Lucas and Osborne (1997) have investigated the conversations of children and adults in botanic gardens, zoos and museums and found that the students within school groups asked more questions and made more statements regarding knowledge than did children in family parties. They credited such a finding to the fact that the format of conversations in informal settings is 'midway between the dominant teacher led dialogue of the classroom, where the teacher asks most of the questions and the situation in homes where the child initiates most dialogue with information-seeking questions' (p. 1053).

3.2 Storytelling and Analogies in Science Teaching and Learning

The communication of scientific ideas to students has been considered as a complex and challenging task which requires 'assigning, developing, or expanding meaning; offering a justification; providing a description; or giving a causal account' (Harlow & Jones, 2004, p. 546). It appears that explaining science is mainly anecdotal, lacking any systematic or thought-out basis (Ogborn, Kress, Martins, & McGillicuddy, 1996). However, research has indicated that storytelling and the use of analogies during instruction is helpful for students to build scientific ideas (Klassen, 2010; Niebert, Marsch, & Treagust, 2012).

Stories are believed to be the primary means by which we make sense of things in our everyday thinking and living (Bruner, 1996). A substantial body of evidence shows that the use of science stories is effective in improving the teaching and learning of science. That is, through a storytelling mode, scientific knowledge, principles and values can be conveyed to students in an accurate, attractive, imaginative and memorable way (Kirchhoff, 2008). However, there is no established basis for evaluating stories other than observing their effect on learning when they are used in the classroom. Noting that there are varying viewpoints on the story form of discourse, to frame the definition of 'storytelling' used in this study, we adopted the notion of the 'literary story' (Klassen, 2009), which highlights the historical and scientific merits as well as the literary merits of the story. Research has suggested that including these stories in science instruction has several benefits, such as: 'making the concepts being taught more memorable, reducing teacher–student distance, assisting in illuminating a point, providing 'reasons for needing to know', stimulating the raising of pertinent questions in the minds of students, and producing explanation-seeking curiosity, of both a historical and scientific type, in students' minds' (Klassen, 2009, p. 417).

Analogies have played an important role in scientific discoveries as the means for scientists to explain fundamentally important concepts (Glynn, 2008). An analogy is the comparison of two similar concepts by pointing out shared characteristics, with the goal of showing that if two things are similar in one way, they are similar in other ways as well (Aubusson, Harrison, & Ritchie, 2006; Coll, France, & Taylor, 2005). In science education, analogies are viewed as effective teaching tools as they 'facilitate understanding the abstract by pointing to similarities in the real world, provide visualization of the abstracts and incite students' interests' (Duit, 1991, p. 414). Although teachers normally acknowledge analogies as being valuable teaching aids, a significant body of research has shown that few are competent in using them during practice (Wilson & Mant, 2011). Moreover, teachers were found to use analogies in a mainly descriptive or explanatory way without any critical considerations through a transmission-reception model of teaching (Oliva, Azcárate, & Navarrete, 2007). In order to address this point and enhance students' understanding of science, teachers have been advised to choose analogues that are familiar to them, with attributes that are precisely identified, and to leave enough time for discussing the comparisons for which the analogy breaks down (Aubusson, Treagust, & Harrison, 2009; Harrison & Treagust, 2006).

References

Aguiar, O. G., Mortimer, E. F., & Scott, P. H. (2010). Learning from and responding to students' questions: The authoritative and dialogic tension. *Journal of Research in Science Teaching, 47*(2), 174–193.

Alexander, R. (2006). *Towards dialogic teaching: Rethinking classroom talk* (3rd ed.). New York: Dialogos.

Alexander, R. (2008). *Essays on pedagogy.* New York: Routledge.

Anderson, L. W., & Krathwohl, D. R. (Eds.). (2001). *A taxonomy for learning, teaching, and assessing: A revision of Bloom's taxonomy of educational objectivities.* New York: Longman.

Ash, D., & Wells, G. (2006). Dialogic inquiry in classroom and museum: Actions, tools and talk. In Z. Bekerman, N. C. Burbules, & D. S. Keller (Eds.), *Learning in places: The informal education reader* (pp. 35–54). New York: Peter Lang.

Atkin, M., & Black, P. (2003). *Inside science education reform: A history of curricular change.* New York: Teachers College Press.

Aubusson, P. J., Harrison, A. G., & Ritchie, S. M. (2006). Metaphor and analogy: Serious thought in science education. In P. J. Aubusson, A. G. Harrison, & S. M. Ritchie (Eds.), *Metaphor and analogy in science education* (pp. 1–10). Dordrecht: Springer.

Aubusson, P. J., Treagust, D., & Harrison, A. G. (2009). Learning and teaching science with analogies and metaphors. In S. M. Ritchie (Ed.), *The world of science education: Handbook of research in Australasia* (pp. 199–216). Rotterdam: Sense Publishers.

Barnes, D. (2008). Exploratory talk for learning. In N. Mercer & S. Hodgkinson (Eds.), *Exploring talk in school* (pp. 1–16). London: SAGE.

Bell, B., & Cowie, B. (2001). *Formative assessment and science education.* Dordrecht, The Netherlands: Kluwer Academic Publishers.

Black, P., & Wiliam, D. (2001). *Inside the blackbox: Raising standards through classroom assessment.* London: King's College London.

Bruner, J. S. (1996). *The culture of education.* Cambridge, MA: Harvard University Press.

Carlsen, W. S. (1991). Questioning in classrooms: A sociolinguistic perspective. *Review of Educational Research, 61*(2), 157–178.

Cazden, C. B. (2001). *Classroom discourse: The language of teaching and learning* (2nd ed.). Portsmouth, NH: Greenwood Press.

Chin, C. (2006). Classroom interaction in science: Teacher questioning and feedback to students' responses. *International Journal of Science Education, 28*(11), 1315–1346.

Chin, C. (2007). Teacher questioning in science classrooms: Approaches that stimulate productive thinking. *Journal of Research in Science Teaching, 44*(6), 815–843.

Chin, C., & Brown, D. E. (2000). Learning in science: A comparison of deep and surface approaches. *Journal of Research in Science Teaching, 37*(2), 109–138.

Chin, C., Brown, D. E., & Bruce, B. C. (2002). Student-generated questions: A meaningful aspect of learning in science. *International Journal of Science Education, 24*(5), 521–549.

Chin, C., & Kayalvizhi, G. (2005). What do pupils think of open science investigations? A study of Singaporean primary 6 pupils. *Educational Research, 47*(1), 107–126.

Chin, C., & Osborne, J. (2008). Students' questions: A potential resource for teaching and learning science. *Studies in Science Education, 44*(1), 1–39.

Coll, R., France, B., & Taylor, I. (2005). The role of models/and analogies in science education: Implications from research. *International Journal of Science Education, 27*(2), 183–198.

Cox-Petersen, A. M., Marsh, D. D., Kisiel, J., & Melber, L. M. (2003). Investigation of guided school tours, student learning, and science reform recommendations at a museum of natural history. *Journal of Research in Science Teaching, 40*(2), 200–218.

Cuccio-Schirripa, S., & Steine, H. E. (2000). Enhancement and analysis of science question level for middle school students. *Journal of Research in Science Teaching, 37*(2), 210–224.

DeWitt, J., & Hohenstein, J. (2010). School trips and classroom lessons: An investigation into teacher-student talk in two settings. *Journal of Research in Science Teaching, 47*(4), 454–473.

DeWitt, J., & Osborne, J. (2007). Supporting teachers on science-focused school trips: Towards an integrated framework of theory and practices. *International Journal of Science Education, 29*(6), 685–710.

Dillon, J. T. (1982). The effect of questions in education and other enterprises. *Journal of Curriculum Studies, 14*(2), 127–152.

Dillon, J. T. (1988). The remedial status of student questioning. *Journal of Curriculum Studies, 20*(3), 197–210.

Dori, Y. J., & Herscovitz, O. (1999). Question-posing capability as an alternative evaluation method: Analysis of an environmental case study. *Journal of Research in Science Teaching, 36*(4), 411–430.

Duit, R. (1991). On the role of analogies and metaphors in learning science. *Science Education, 75*(6), 649–672.

Edwards, D., & Mercer, N. (1987). *Common knowledge: The development of understanding in the classroom*. London: Routledge.

Edwards, D., & Westgate, D. (1994). *Investigating classroom talk*. London: Falmer.

Eggleston, J. F., Galton, M., & Jones, M. (1976). *Processes and products of science teaching*. London: Macmillan.

Glynn, S. M. (2008). Making science concepts meaningful to students: Teaching with analogies. In S. Mikelskis-Seifert, U. Ringelband, & M. Brückmann (Eds.), *Four decades of research in science education: From curriculum development to quality improvement* (pp. 113–125). Müster: Waxmann.

Good, T. L., Slavings, R. L., Harel, K. H., & Emerson, H. (1987). Student passivity: A study of question asking in K-12 classrooms. *Sociology of Education, 60*(3), 181–199.

Graesser, A. C., & Person, N. K. (1994). Question asking during tutoring. *American Educational Research Journal, 31*(1), 104–137.

Griffin, J. M., & Symington, D. (1997). Moving from task-oriented to learning-oriented strategies on school excursions to museums. *Science Education, 81*(6), 763–779.

Hardman, F. (2008). Opening-up classroom discourse: The importance of teacher feedback. In N. Mercer & S. Hodgkinson (Eds.), *Exploring talk in school* (pp. 131–150). London: SAGE.

Harlow, A., & Jones, A. (2004). Why students answer TIMSS science test items the way they do. *Research in Science Education, 34*(2), 221–238.

Harper, K. A., Etkina, E., & Lin, Y. (2003). Encouraging and analysing student questions in a large physics course: Meaningful patterns for instructors. *Journal of Research in Science Teaching, 40*(8), 776–791.

Harrison, A. G., & Treagust, D. F. (2006). Teaching and learning with analogies: Friend or foe? In P. J. Aubusson, A. G. Harrison, & S. M. Ritchie (Eds.), *Metaphor and analogy in science education* (pp. 11–24). Dordrecht: Springer.

King, A. (1990). Enhancing peer interaction and learning in the classroom through reciprocal questioning. *American Educational Research Journal, 27*(4), 664–687.

King, H. (2009). *Supporting natural history enquiry in an informal setting: A study of museum explainer practice* (Unpublished doctoral dissertation). King's College London, London, UK.

Kirchhoff, A. (2008). Weaving in the story of science. *Science Teacher, 75*(3), 33–37.

Klassen, S. (2009). The construction and analysis of a science story: A proposed methodology. *Science & Education, 18*(3–4), 401–423.

Klassen, S. (2010). The relation of story structure to a model of conceptual change in science learning. *Science & Education, 19*(3), 305–317.

Kozulin, A. (2003). Psychological tools and mediated learning. In A. Kozulin, B. Gindis, V. S. Ageyev, & S. M. Miller (Eds.), *Vygotsky's educational theory in cultural context* (pp. 15–38). Cambridge, MA: Cambridge University Press.

Lemke, J. L. (1990). *Talking science: Language, learning and values.* Norwood, NJ: Ablex Publishing.

Marbach-Ad, G., & Sokolove, P. G. (2000). Can undergraduate biology students learn to ask higher-level questions? *Journal of Research in Science Teaching, 37*(8), 854–870.

Maskill, R., & Pedrosa de Jesus, H. (1997). Pupils' questions, alternative frameworks and the design of science teaching. *International Journal of Science Education, 19*(7), 781–799.

Mehan, H. (1979). *Learning lessons: Social organisation in the classroom.* Cambridge, MA: Harvard University Press.

Mercer, N. (1995). *The guided construction of knowledge: Talk amongst teachers and learners.* Clevedon: Multilingual Matters.

Mercer, N. (2000). *Words and minds: How we use language to think together.* London: Routledge.

Mercer, N. (2008). The seeds of time: Why classroom dialogue needs a temporal analysis. *Journal of the Learning Sciences, 17*(1), 33–59.

Mercer, N., & Howe, C. (2012). Explaining the dialogic processes of teaching and learning: The value and potential of sociocultural theory. *Learning, Culture and Social Interaction, 1*(1), 12–21.

Mercer, N., & Littleton, K. (2007). *Dialogue and the development of children's thinking: A sociocultural approach.* London: Routledge.

Nassaji, H., & Wells, G. (2000). What's the use of triadic dialogue?: An investigation of teacher-student interaction. *Applied Linguistics, 21*(3), 376–406.

Ng-Cheong, J., & Chin, C. (2009). *Questioning as a learning strategy in primary science.* Paper presented at the International Science Education Conference 2009, National Institute of Education, Singapore.

Niebert, K., Marsch, S., & Treagust, D. F. (2012). Understanding needs embodiment: A theory-guided reanalysis of the role of metaphors and analogies in understanding science. *Science Education, 96*(5), 849–877.

Nystrand, M. (1997). *Opening dialogue: Understanding the dynamics of language and learning in the English classroom.* New York: Teachers College Press.

Nystrand, M., Wu, L. L., Gamoran, A., Zeiser, S., & Long, D. A. (2003). Questions in time investigating the structure and dynamics of unfolding classroom discourse. *Discourse Processes, 35*(2), 135–198.

O'Connor, M. C., & Michaels, S. (1996). Shifting participant frameworks: Orchestrating thinking practices in group discussion. In D. Hicks (Ed.), *Discourse, learning, and schools* (pp. 63–103). Cambridge, MA: Cambridge University Press.

Ogborn, J., Kress, G., Martins, I., & McGillicuddy, K. (1996). *Explaining science in the classroom.* Buckingham: Open University Press.

Oliva, J. M., Azcárate, P., & Navarrete, A. (2007). Teaching models in the use of analogies as a resource in the science classroom. *International Journal of Science Education, 29*(1), 45–66.

Osborne, J. (2005). *The challenge of materials gallery: A discourse based cognitive analysis.* Paper presented at the Annual Conference of the National Association for Research in Science Teaching (NARST), Dallas, TX.

Pedrosa de Jesus, H., Almeida, P., & Watts, M. (2004). Questioning styles and students' learning: Four case studies. *Educational Psychology, 24*(4), 531–548.

Pedrosa de Jesus, H., Teixeira-Dias, J., & Watts, M. (2003). Questions of chemistry. *International Journal of Science Education, 25*(8), 1015–1034.

Redfield, D. L., & Rousseau, E. W. (1981). A meta-analysis of experimental research on teacher questioning behaviour. *Review of Educational Research, 51*(2), 237–245.

Rogoff, B. (1990). *Apprenticeship in thinking: Cognitive development in social context.* Oxford: Oxford University Press.

Rojas-Drummond, S. (2000). Guided participation, discourse and the construction of knowledge in Mexican classrooms. In B. Cowie & G. Van der Aalsvoort (Eds.), *Social interaction in learning and construction: The meaning of discourse for the construction of knowledge* (pp. 193–213). Oxford: Pergamon.

Rojas-Drummond, S., & Mercer, N. (2003). Scaffolding the development of effective collaboration and learning. *International Journal of Educational Research, 39*(1–2), 99–111.

Rojas-Drummond, S., Mercer, N., & Dabrowski, E. (2001). Collaboration, scaffolding and the promotion of problem solving strategies in Mexican pre-schoolers. *European Journal of Psychology of Education, 16*(2), 176–196.

Scardamalia, M., & Bereiter, C. (1992). Text-based and knowledge-based questioning by children. *Cognition and Instruction, 9*(3), 177–199.

Scott, P. H. (1998). Teacher talk and meaning making in science classrooms: A Vygotskian analysis and review. *Studies in Science Education, 32*(1), 45–80.

Scott, P. H., Mortimer, E. F., & Aguiar, O. G. (2006). The tension between authoritative and dialogic discourse: A fundamental characteristic of meaning making interactions in high school science lessons. *Science Education, 90*(4), 605–631.

Sinclair, J. M., & Coulthard, M. (1975). *Towards an analysis of discourse.* London: Oxford University Press.

Smith, F., Hardman, F., Wall, K., & Mroz, M. (2004). Interactive whole class teaching in the National Literacy and Numeracy Strategies. *British Educational Research Journal, 30*(3), 395–411.

Tal, T. (2012). Imitating the family visit: Small-group exploration in an ecological garden. In E. Davidsson & A. Jakobsson (Eds.), *Understanding interactions at science centers and museums: Approaching sociocultural perspectives* (pp. 193–206). Rotterdam: Sense.

Tal, T., & Morag, O. (2007). School visits to natural history museums: Teaching or enriching? *Journal of Research in Science Teaching, 44*(5), 747–769.

Tunnicliffe, S. D., Lucas, A. M., & Osborne, J. (1997). School visits to zoos and museums: A missed educational opportunity? *International Journal of Science Education, 19*(9), 1039–1056.

van Zee, E. H., & Minstrell, J. (1997). Using questioning to guide student thinking. *The Journal of the Learning Sciences, 6*(2), 227–269.

Vygotsky, L. S. (1978). *Mind in society.* Cambridge, MA: Harvard University Press.

Vygotsky, L. S. (1987). Thinking and speech (N. Minick, Trans.). In R. W. Rieber, A. S. Carton & J. S. Bruner (Eds.), *The collected works of L.S. Vygotsky: Problems of general psychology* (Vol. 1). London: Plenum.

Watts, M., Gould, G., & Alsop, S. (1997). Questions of understanding: Categorising pupils' questions in science. *School Science Review, 79*(286), 57–63.

Wells, G., & Arauz, R. M. (2006). Dialogue in the classroom. *Journal of the Learning Sciences, 15*(3), 379–428.

Wertsch, J. V. (1991). *Voices of the mind: A sociocultural approach to mediated action.* London: Harvester Wheatsheaf.

Wertsch, J. V. (1998). *Mind as action.* Oxford: Oxford University Press.

Wilson, H., & Mant, J. (2011). What makes an exemplary teacher of science? The teachers' perspectives. *School Science Review, 93*(343), 115–119.

Chapter 4
Research Design and Methodologies

This study was set out to investigate two aspects of BGEs' talk during the guided school visits: the functions of their talk and how their talk mediates students' learning of science. To address these matters, this study followed a qualitative approach, which facilitates the enquiry into the phenomena taking place in the natural setting (Bodgan & Biklen, 2007; Creswell, 2008).

4.1 A Case Study Approach

To explore, describe and explain the nature of BGEs' pedagogy in supporting and enriching visiting schoolchildren's learning about ecological science are both purposes and strategies of this study. In order to generate robust outcomes that allow for theory development, case study was selected as the research approach.

Gillham (2000) defines the word 'case' as: 'a unit of human activity embedded in the real world; which can only be studied or understood in context; which exists in the here and now; that merges in with its context so that precise boundaries are difficult to draw' (p. 1). Moreover, a case study is a strategy for carrying out research which involves an investigation of a contemporary phenomenon within its real-life context, using multiple sources of evidence (Patton, 2002; Yin, 2003). In qualitative case studies, researchers can investigate a bounded system (single-case study) or several bounded systems (multiple-case study) over time (Creswell, 2008). However, Miles and Huberman (1994) argue that it is advantageous if more than one case can be researched, because it enhances the understanding gained: 'multiple cases offer the researcher an even deeper understanding of processes and outcomes of cases, the chance to test (not just develop) hypotheses and a good picture of locally grounded causality' (p. 26). Overall, the evidence from a case study is often considered more compelling, and thus it is regarded as being more robust (Yin, 2003). Therefore, this study is conducted following a case study approach in order to obtain

© Springer Science+Business Media Singapore 2015
J. Zhai, *Teaching Science in Out-of-School Settings*,
DOI 10.1007/978-981-287-591-4_4

extensive data from the cases. In this study, each individual BGE's teaching practices is considered as a case so as to look in-depth into the BGEs' pedagogies for effective learning.

4.2   Research Strategies

Getting access to the research sites and participants was a major challenge in this study, because as Osborne and Dillon (2007) have highlighted, research exploring teaching and learning in informal contexts is more complex and complicated than that sited in formal settings. In this section the decisions and actions taken regarding recruiting participants, brief descriptions of the research sites and explanations regarding how I built rapport with the BGEs are presented. The research was carried out in three phases: the initial fieldwork, a pilot study and the main study.

The initial fieldwork phase commenced in June 2008. The purpose of this stage was to become familiar with the settings and develop a rapport with the participating BGEs. During this period, once permission had been obtained from the botanic gardens, I observed the school visits and took notes and thus did not need to obtain permission from the visiting school groups. In brief, each BGE was shadowed for approximately 1 week and some of their frequently used teaching strategies were noted down. When I left the field, much time was spent on developing a research framework, which included the foci for use during the observations, how to assess the children' visit experiences and strategies for capturing the BGEs' reflection on their practice.

With the research framework developed from the initial fieldwork, I went back to the gardens to carry out the pilot leading to the main study. Letters containing information about this research and consent forms were sent to the schoolteachers, students and their parents using contact information provided by the BGEs. When the schoolchildren came to the gardens, the visits were video recorded so that they were available for use during the development of the analytical framework at a later stage of this research. Unfortunately, the camcorder that was used had either run out of batteries by the middle of the observation or the microphone connected to the camera failed to work and consequently, only a small amount of data in this medium was available for analysis. These events led to the updating of the camcorder model and preparation of extra batteries and memory cards for use during the main phase of the study. In order to investigate each student's perception of the botanic garden and the nature of the visiting experience in an in-depth manner, for the pilot I attempted to hold interviews with the schoolchildren, both before and after their visits, by attending their school in person. However, the teacher only permitted 10 min for each interview, which in fact was too short a time in which to elicit in-depth information. Moreover, since time and budget restrictions did not allow for travel to each visiting school and thus for the main study, it was decided to rely on a student survey in order to assess their experiences regarding the botanic garden visit.

After reflecting on these experiences during the pilot phase, the key change was to revise the data collection scheme for the main study so as to meet the realities encountered in the field. Other changes included obtaining a high-quality microphone and using both audio and video recorders for backup purposes. In sum, the initial fieldwork phase helped to build a rapport with the BGEs and raise my awareness of their teaching styles, whilst the pilot phase made me aware of technical and methodological issues that could impact on the effective execution of the main study.

4.2.1 Description of the Sites and Participating BGEs

Once the BGEs had been recruited for this study, I asked the managers in the botanic gardens to help with contacting the prospective visiting school groups, so that they could be invited to participate in the research project. All the details regarding the research were laid out in a bundle of documents that included: the research proposal, an information letter for the students and a consent form for the parents to sign, which was forwarded to the schoolteachers who were planning to visit the garden during the 2008/2009 academic year. This section contains thumbnail sketches of the three participating botanic gardens, covering some general data about the sites and their education programmes for school groups.

Garden A is located in the metropolitan area of City X. Although relatively small in size, it holds approximately 5,000 taxa, and the focus of the collection is on medicinal plants and those of ethnobotanical interest, as well as rare and endangered species. During four decades of development, the education department in Garden A has designed a set of education programmes for visiting school groups, ranging from KS1 to KS4. Mark was the only full-time employed BGE and he was responsible for teaching school visits, providing local schools with continuing professional development training and organizing educational events for the visiting general public.

Garden B is located in City Y and covers 15 acres. Besides glasshouses containing various tropical, citrus and succulent plants, the site also offers an adventure playground and children's discovery garden, both of which provide entertainment for younger visitors, many of whom are also enthralled by the additional feature of a bird collection. The education centre in Garden B fell under two administrations, namely, those of the local city council and its own internal education department. Simon and Debbie were the full-time BGEs employed by the two administrations, respectively.

Garden C is a 40-acre garden located in City Z, a university-based town. The botanic garden is an outreach branch of the local university and was established as a teaching and research resource more than a hundred years ago. It features a collection of over 10,000 labelled plant species in beautifully landscaped settings, including: a rock garden, a lake, glasshouses, a winter garden, a woodland walk and national collections. According to its official statistics, the garden attracts around

Table 4.1 Participating BGEs

BGE	Education background	Working as a BGE (years)	Other teaching/working experience
Mark (Garden A)	BSc in ecology	15	Outdoor educator for 15 years
Simon (Garden B)	BSc in applied science	7	Primary teacher for 15 years
	PGCE in secondary science		Outdoor educator for 3 years
Debbie (Garden B)	BA with QTS (focused on environmental studies)	5	LEA education officer for 4 years
			Outdoor educator for 6 years
Julia (Garden C)	PhD in plant physiology	3.5	Landscape designer for 3 years
	PGCE in primary teaching		Primary teacher for 4 years
			Programme officer in special school for 2 years

100,000 visitors each year, including 8,000 students on educational visits. As one part of its mission, Garden C provides educational visits to schoolchildren of different ages. Julia (a BGE) was the head of the education department, and she alone was in charge of school visits, whilst the other two staff were responsible for adult education programmes.

The rationale for selecting the BGEs who were to participate in this study was largely determined by the situation regarding the arrangement for teaching responsibilities in each of the three research sites. That is, only those BGEs who taught visiting school groups were considered to be suitable participants. Table 4.1 contains information about the participating BGEs.

4.2.2 Building Rapport Throughout the Research

To collect robust data, as LeCompte and Schensul (1999) have suggested, it is important for the qualitative researcher to develop a close involvement with the research participants comprising the building of trust and in some cases developing friendships. Therefore, it was deemed appropriate to observe each participating BGE for 2 weeks, whilst waiting for the feedback from the schools that had been invited to participate in the research. To establish a working relationship, in addition to observing the school visits, I helped the BGEs prepare their teaching materials and set up venues for the learning activities for the school groups. Through involvement in these tasks I became familiar with the specific botanic garden environment and the way in which the BGEs organized and delivered the guided visits. Moreover, through such processes the BGEs and I gradually built trusting relationships, which facilitated the informal communication of rich information regarding the BGEs, such as, their backgrounds, perceptions regarding teaching and learning and the difficulties they encountered in their work.

4.3 Data Collection Methods

The power of the case study approach is that it allows the researcher to use a variety of sources, data and research methods, as part of the overall investigation (Denscombe, 2003; Stake, 1995; Yin, 2003). With regard to this, in the current study observations, interviews and a student survey were used to obtain different sets of data, which afforded the opportunity for triangulating the different sets of information so as to produce a valid analysis of the BGEs' pedagogical practices. In the following section, each of the data collection methods is addressed.

4.3.1 Observations

Observation is 'the systematic description of events, behaviours and artefacts in the social setting chosen for study' (Marshall & Rossman, 2006, p. 96), and through this process the researcher can look at what is taking place in situ. As the focus here rests in exploring the botanic garden as a teaching and learning setting, the pedagogical approaches that BGEs employ to engage and support children's learning and their teaching behaviours, it was decided that observation was to be the primary strategy for collecting data regarding these phenomena.

In the social sciences, observational strategies are classified according to the extent to which the observer is a participant in the setting being investigated. Gold (1958) depicts the extent of participation as a continuum, ranging from complete immersion in the setting, with the researcher as a full participant, and at the other extreme, complete separation from the setting, where the researcher is as a spectator. Between these two poles, the researcher may act as participant-as-observer or observer-as-participant depending on the extent of his/her participation. Participant-as-observer refers to the situation wherein the researcher's observing activities are subordinate to the role as a participant, whereas observer-as-participant indicates that the researcher's involvement is secondary to his/her role as information gatherer. Further, Adler and Adler (1994) termed the participant-as-observer as having an active membership role and observer-as-participant as having a peripheral membership role.

The researcher's observational role is not fixed and the extent of participation can, in certain circumstances, change over time. With regard to this, in the observations for the current study, I tended to stay at the back of the classes, especially during whole class instruction sessions, in the role of a complete observer in order to minimize any potential disruption to the natural teaching settings. Complete observation, in which the researcher only takes an observation role, has been criticized for its minimal reactivity (Cohen, Manion, & Morrison, 2000), but its advantage and the reason for using it in this research is that it investigates naturalistic phenomena as they happen.

Recent advances in digital video technology and video analysis can facilitate a deeper insight into the interplay of teaching and learning processes, when they are developed in both formal and informal contexts (Barron, 2007; Derry, 2007). That is, the effective use of video can assist researchers in creating descriptive, explanatory or expository accounts of these processes (Goldman, Pea, Barron, & Derry, 2007). In light of the opportunities that this technology can offer, video recording was selected as the primary data collection tool for this fieldwork, because it has clear practical advantages over note-taking in the field, for example, more detail is recorded which can be revisited as many times as is necessary, at a later date. In other words, first, through collecting videos of events, the researcher can 'create more closely grounded stories that include the full range of gestural, auditory and contextual subtlety in the thick description of the event' (Goldman, 2007, p. 4), and second, the raw video data can be reanalysed by other scholars to check the validity of the categories or theories developed from the initial researcher's analysis (Shrum, Duque, & Brown, 2005).

In spite of there being many advantages associated with the use of technology, as described above, the presence of a video camera and/or a voice recorder may affect participants' behaviours. However, some researchers who have used videoing as a data collection tool in classrooms have found that participants often notice the equipment at the very beginning of the recording but tend to forgot about being recorded as long as they are fully engaged in the class activities (Barron, 2007; Jacobs, Kawanaka, & Stigler, 1999; Stigler, Gonzales, Kawanka, Knoll, & Serrano, 1999). Turning to the current study, the experience of making recordings during the pilot phase and the later main study was largely consistent with these scholars' findings. I agree with Barron's (2000) argument that:

> Although it is possible that the video camera may have influenced student behaviour, it is difficult to predict in which direction. Being recorded could as easily have been distracting as facilitating with respect to the attention of the student participants. (p. 397)

The observations that comprised the main study in the field were conducted from April to June 2009 (see Table 4.2 for the detailed schedule). For these observations each school trip was recorded, using a single, consumer-grade video camcorder. The aim was to capture what a typical student tended to focus on during the guided visit, which usually was the BGE during whole class teaching sessions. Hall (2007) suggests that the researcher who uses a single camera should make continuous recording and minimize panning and zooming. Thus, before the recording, the BGEs and students were informed that the purpose of the study was to record typical guided school visits and that they should act as they would do normally. The videoing started as soon as the BGE started teaching and continued until the end of the lesson and breaks were taken only when it was necessary to change the battery and/or the memory card in the camera. Regarding where to point the camera, both stationary and hand-held recordings were included (Barron, 2007). As discovered during the initial fieldwork and pilot study experiences, the students' activities inside the classroom were mainly seated, and thus, to capture these, the camera was positioned on a tripod at the back of the class. In contrast, for the sessions

Table 4.2 Observation schedule for all the recorded visits

BGE	School groups	Grade	Number of students	Topic (s)
Mark	BS	Y3	19	Habitats and plant adaptations
	FP	Y5	39	Habitats and plant adaptations
Simon	SB	Y3	29	Plants and growing
	NP	Y3	19	Plants and growing
Debbie	GS	Y8	20	Amazon Rainforests and rainforest instruments
	WM	Y8	20	Amazon Rainforests
Julia	SF	Y5	11	Habitats
	UW	Y5	10	Plant adaptations and seeds and Darwin

conducted outside, it was more convenient to hold the camera, especially when the groups were in the narrow glasshouses, which quickly became very crowded. Although the digital camcorder was programmed to keep the target participants in its picture frame, the inbuilt microphone often failed to capture their voices when they were at a distance from the equipment. Therefore, in order to record all the BGEs' discourses, each of the BGEs wore a clip-on microphone connected to a digital voice recorder, and these audio data files were used later for transcribing any conversations that were inaudible from the video recordings.

Simon and Debbie, the BGEs in Garden B, did not want to be filmed when they were teaching, but they allowed an audio recording of the visits and for field notes to be made. Thus, an observation schedule, based on the work of Choi (2007), was developed to record the school visits in this particular setting. The fieldwork in this garden involved making time-indexed note as the audio recording was in progress, and when the audio data was transcribed, these field notes were used to annotate the transcript with details regarding the setting along with the non-verbal behaviours of the participants.

4.3.2 Interviews

In order to investigate their understanding of pedagogy and their teaching experiences, all the participating BGEs were formally interviewed twice. The interview data containing their perspectives on teaching and learning as well as their reflections on the observed lessons were compared with what was actually observed and as such, served to triangulate the data. Moreover, this procedure allowed for a comparison between the BGEs preferred practice and the observed reality.

Patton (2002) proposes that 'observations provide a check on what is reported in interviews; interviews, on the other hand, permit the observer to go beyond external behaviour to explore the internal states of persons who have been observed' (p. 306). With respect to this, it is accepted that interviews bridge the gap in data collection that observations are unable to cover, that is, they help in discovering respondents' feelings, thoughts and intentions. Thus, through carrying out interviews, the researcher can seek to access 'people's perceptions, meanings, definitions of situations and constructions of reality' (Punch, 2005, p. 168). According to the extant literature, interviews have been used as an important data collection tool within the limited research that has examined the pedagogical practices of informal educators. Interviews with participants have been used to prompt subjects to reflect on their teaching practice and, thus, have enriched researchers' primary analyses of observed behaviour gathered in their fieldwork. For example, through carrying out interviews prior to and after lesson observations, Tran (2006) found that such educators were consciously aware of their actions and were able to provide reasoned accounts of their practice. Similarly, in the analysis of educator-student interactions during a museum educator-led workshop, King (2009) relied on interviews for further probing of the pedagogical beliefs and ontological understandings held by the educators. Through these interviews, she found that many aspects of the educator practices were only tacitly understood and that for these educators, notions of educational theory were often misconceived. Turning to the current study, interviews were chosen as a key means of data collection in order to elicit more detailed and complex information regarding the BGEs' beliefs and intentions than that data available from solely relying on making observations of their practices.

The interview style adopted was semi-structured, because this method provides the interviewer with 'considerable flexibility over the range and order of questions within a loosely defined framework' (Wellington, 2000, p. 74). For instance, under this approach it was possible to prompt and probe the participants' interpretations and explanations of their experiences of teaching on school visits. The observations carried out before the main study provided the researcher with the opportunity to develop a rapport with the participants, which contributed to the creation of a friendly and comfortable atmosphere during the formal interviews. There were two rounds, one prior to the main study field observations and the second after. All the interviews took place on the research sites and were audio-taped for later transcription, with each ranging from 20 to 50 min, depending on the interviewee's availability.

The first round interviews took place between April and June 2009, with an interview protocol being developed with the aim of ensuring that the nature of each BGE's beliefs was fully explored. More specifically, the goal of these questions was to obtain insight into the participating BGEs' professional backgrounds, their responsibilities for school visits and their perceptions regarding teaching and learning in informal contexts, particularly in botanic gardens. Although the interviews were semi-structured, some structured, questionnaire-like questions were asked when eliciting the participants' basic demographic information.

A few weeks after completing the field observations, the second round interviews were conducted during June and July 2009. Before returning to the research sites to interview the BGEs, all the recorded observation data was processed using editing software and from this some interesting clips were chosen at random so as to serve as illustrations of particular pedagogic practices, such as those involving:

whole class teaching, student group work and guided tours. A few of these excerpts, in the form of video clips and audio transcriptions, were played to the BGEs during the interviews, in order to prompt their memories regarding the observed visits. After watching, listening to or reading out the edited transcripts, I started the interview by asking 'what do you think about these lessons?' and then each BGE reflected on his/her teaching, taking into account what they saw on the screen and what they could recall. Although the interview protocol was always available for guidance and reference, the key thrust of the questioning was to find out what the respondents thought about their teaching, whether they felt that they had achieved their objectives, what they felt about their decisions and actions and why they had used particular strategies in the observed teaching sessions.

During both rounds of interviews, a record of the interviewee's physical position, disposition and attitude was noted, because according to Wellington (2000), the combination of note-taking and tape-recording during an interview enriches the 'texture of reality' (p. 85) that can be captured. Having returned from the research sites, all the interviews were transcribed with additional annotations indicating the interviewee's gestures and any moments of silence. In order to enhance the validity of the interview data, the transcripts were sent back to the BGEs via email for appraisal and checking. However, none of the participants replied to me, offering any corrections or amendments to their interview transcripts.

4.3.3   Impression Sheets

Yin (2003) suggests that the case study should include multiple data sources in a 'triangulation fashion' (p. 14), as this facilitates validation, because of the opportunities it provides for cross verification. The last data collection tool employed in the fieldwork was the impression sheet. These were distributed to the school parties with the purpose of evaluating the students' experiences regarding the botanic garden visits and providing data that could be triangulated with information gathered from the BGEs' responses in the interviews and the observations.

The aim of asking students to complete the impression sheets was to evaluate their visiting experiences. This information could have been gathered by carrying out interviews after the visits, but for a number of reasons, the impression sheets were deemed to be most appropriate for this fieldwork. To begin with, the distribution and administration of this tool is, in the main, more economical and time efficient than holding interviews. Moreover, as I am a speaker of English as a foreign language and have relatively limited experience of working with young children, the application of straightforward impression sheets offered a more practical and reliable means through which to collect data than face-to-face interviews. Furthermore, if interviews were to be held with the students on the site immediately after the BGE-guided visit, their visiting experience could have been adversely affected, because they could have lost the time for exploring other parts of the botanic garden by themselves or with their schoolteacher.

In order to minimize my impact on the students' visit and yet still be able to assess their experiences during the trip, a questionnaire was designed following Cohen et al.'s (2000) advice that 'qualitative, less structured, word-based and open-ended questionnaires' can be helpful to 'capture the specificity of a particular situation' (pp. 247–248). Thus, an impression sheet, a form of questionnaire that engaged the children and encouraged them to give both written and/or drawn responses to the questions, was developed for this study. To ensure that the students could understand all the questions set on the sheet, the Bristol norms for age of acquisition of words (Gilhooly & Logie, 1980; Stadthagen-Gonzalez & Davis, 2006) were enlisted. There were eight questions on the impression sheet aimed at gathering student attitudes towards: the learning activities during their visit, the subject knowledge they had acquired and their perception of the BGE instruction. All these questions were open ended in nature, which increased the likelihood that the students would 'write [and draw] a free response in their own terms, to explain and qualify their responses and avoid limitations of pre-set categories of response' (Cohen et al., 2000, p. 248).

The impression sheets were given to the schoolteachers, who were in charge of each visiting group before they left the botanic garden to return to school, with the request being made that they distribute the sheets to students and return them within 1 week after the visit. To ensure a reasonable return rate, the teachers were provided with a stamped, addressed envelope, but four school groups out of eight did not comply and no sheets were returned (Table 4.3).

From the above table, it can be seen that the return rate for those schools that sent back impression sheets is impressively high, especially from the UW School group. Regarding this, it was impossible to tell whether the students were coerced into completing the sheets or filled them in voluntarily, but when they were given to the schoolteachers, it was suggested that they should take part in the survey of their own volition.

Table 4.3 The return rate of the impression sheets

BGE	School group	Number of impression sheets distributed	Number of impression sheets returned
Mark	BS	19	0
	FP	39	36
Simon	SB	29	0
	NP	19	0
Debbie	GS	40	38
	WM	40	0
Julia	SF	39	34
	UW	20	20

4.4   Data Analysis

As Firestone (1993) suggested, 'multi-case studies can use the logic of replication and comparison to strengthen conclusions drawn in single sites and provide evidence for both their broader utility and the conditions under which they hold' (p. 22). Therefore, here four participating BGEs' teaching practices are presented as individual cases guided by two research questions:

(a) What is the structure of a BGE-guided school trip to a botanic garden?
(b) How do the BGEs interact with students in terms of engaging and supporting their learning?

In order to address the above questions, the data collected from observations, interviews and a survey were analysed, and four distinct areas for the subsequent investigation emerged: the structure of the guided visit, the BGE's perception of teaching and learning, BGE-student interaction and the students' views about their visiting experience. Next, according to these identified areas, there is explanation of how the raw data were processed and coded into themes so as to be able to address the research questions as well as for subsequent theory development.

4.4.1   The Analysis of Discourse Data

As discussed in Chap. 3, the sociocultural perspective of learning highlights the role of discourse in knowledge construction, and the discourse between children and adults is considered as an important lens through which teaching and learning can be assessed in different contexts (DeWitt & Hohenstein, 2010; Edwards & Mercer, 1987; King, 2009; Osborne, Erduran, & Simmon, 2004). In order to investigate the nature of BGE-student interaction during the visit, the transcripts were analysed according to the flow of power in the discourse. In this context, the extent to which the lesson discourse is initiated by students, either sharing personal experience or posing questions for explanation, indicates different levels of students' engagement in the learning process.

After all the video or audio discourse data had been transcribed verbatim, the utterances were demarcated into real units of speech communication, because according to Bakhtin (1986):

> Speech can exist in reality only in the form of concrete utterances of individual speaking people, speech subjects. Speech is always cast in the form of an utterance belonging to a particular speaking subject and outside this form it cannot exist. (p. 71)

With respect to this, in the Third International Mathematics and Science Study (TIMSS), utterance was used as the smallest unit of analysis for describing classroom discourse (Jacobs et al., 2003). Stigler et al. (1999) define an utterance as 'a sentence or phrase that serves a single goal or function', adding that they frequently 'correspond to a single turn in a classroom conversation' (p. 32). In this study, an utterance was adopted as the analytical unit to characterize the interactions between the BGEs and the students.

4.4.1.1   Exchange Structures: Triadic and Atypical Triadic

In order to examine the pattern of BGE-student interaction, the context of what was said, such as previous utterances, concurrent activities and tone of voice, was taken into consideration. This provides important clues for identifying the exchange structure in terms of the initiator of discourse, which in turn helps to identify the authority and hence where the power lies during the discussion. The exchange structures were coded as triadic and atypical triadic. The triadic dialogue in educational settings refers to the I-R-F teacher-student talk in the classroom, where the teacher initiates, the student responds and teacher gives feedback (Mehan, 1979; Sinclair & Coulthard, 1975). With respect to this, a significant body of research has found that triadic dialogue predominates in classroom discourse, with the teacher controlling the communication (Cazden, 2001; Lemke, 1990). Through the field observations, it became apparent that triadic discourse frequently occurred during the BGE-guided visits, and consequently the exchanges were first coded as: 'initiate' (I), 'response' (R) and 'feedback' (F).

The pattern of adult-child interaction in informal settings, such as museums, however, is often different from that in school classrooms, often taking the form of atypical triadic dialogue. In this, the children usually open exchanges by attracting attention to the exhibits by 'using ostensive language, 'Look!', or asking the identity of a specimen using the question 'What is it?' or another similar question' (Tunnicliffe, Lucas, & Osborne, 1997, pp. 1502–1503). In such atypical triadic discourse, the child is the initiator of the exchanges and not the adult. In their research on teacher-student talk in classrooms and on museum visits, DeWitt and Hohenstein (2010) coded student-initiated talk as a 'volunteering statement', in which a student volunteers or shares information, suggests actions, contradicts or corrects the teacher or initiates a new topic of conversation. Such student volunteering statements in the exchanges indicates greater symmetry in teacher-student interactions, because the latter are allowed to offer their points of view (Scott, Mortimer, & Aguiar, 2006). In this study, all the statements and questions initiated by the visiting schoolchildren were coded as 'volunteering talk' (VOL).

4.4.1.2   The Functions of Follow-up Moves

As discussed in Chap. 3, the teacher's follow-up moves play a significant role in directing the development of classroom conversation (Hardman, 2008; Nystrand, Wu, Gamoran, Zeiser, & Long, 2003; Wells & Arauz, 2006). For this research, the frameworks used in other research on educator's follow-up moves were synthesized to help shape the data analysis (see Table 4.4).

Table 4.4   The frameworks used to examine educator follow-up moves

Coding schema used by scholars	Sinclair and Coulthard (1975)	Edwards and Mercer (1987)	O'Connor and Michaels (1996)	Scott (1998)	King (2009)
Maintain					Withhold answer to avoid answering a direct question
Insert	Direct instruction				Explain
Elicit	Cue elicitation, prompt	Cued elicitation		Ask a rhetorical question	Cue to describe, relate, speculate
Evaluate	Accept, evaluate	Confirmation		Accept a student's response	Selects one particular response either overtly or implicitly
Repeat		Repetition to draw whole class's attention to an answer	Rebroadcast utterance to reach a wider audience	Repeat, rephrase	Repeat a particular response to draw significance to it and thus share the utterance with the wider class
Revoice	Comment: the process of exemplifying, expanding, justifying and providing additional information	Reformulation: a revised/tidied up version of what was said Elaborations: expand on or explain the significance of a cryptic statement to rest of class	Revoice to emphasize particular content or structure	Reformulate a question with a particular intonation	Revoice: reformulating the response or rephrasing the vocabulary

Drawing on the literature regarding teacher talk and the trial analysis of the pilot phase data, the BGEs' follow-up moves were coded into: 'insert', 'elicit', 'evaluate', 'repeat' and 'revoice'. 'Insert' refers to the instances when an educator gives information or answers to students as a follow-up to what they have said and when it is used; it serves to support Lobato, Clarke and Ellis' (2005) claim that teachers cannot avoid 'telling'. 'Elicit' involves trying to seek the student's ideas by contributing something new, which is similar to Edwards and Mercer's (1987) 'cued elicitation'. 'Evaluate' refers to the educator's judgement on the students' responses. The 'repeat' move is when the educator's voice draws the students' attention to a

Table 4.5 Coding schema for the BGEs' follow-up moves

Follow-up move (code)	Features/examples
Insert (Int)	The BGE adds something in response to the student's contribution. The BGE can: elaborate up on it, correct it, suggest something, make a link, etc.
Elicit (Ect)	Whilst following up a contribution, the BGE tries to elicit something new from the student or other students. The BGE elicits additional information or a new but related idea to take the lesson forward. Elicit moves often, but not always, narrow the contributions
Maintain (Mnt)	The BGE maintains the contribution in the public realm for further consideration. The BGE can repeat the idea, ask others for comment or merely indicate that the student should continue talking
Evaluate (Evl)	The BGE makes an evaluative judgement regarding a student's contribution, for example, 'good', 'well done', 'excellent', 'brilliant', 'wrong', etc.
Repeat (Rpt)	The BGE repeats what a student has said in the form of a statement
Revoice (Rev)	The BGE reformulates a student's discourse to make it clear for the rest of the class
Confirm (Cfm)	The BGE confirms that she/he has heard the learner correctly. There should be some evidence that the BGE is not sure what she/he has heard from the learner; otherwise it could be a press situation
Press (Prs)	The BGE pushes or probes the student for more on his/her idea, to clarify, justify or explain something more clearly to the rest of the class. The BGE does this by asking the student to explain more why she/he is correct or by asking a specific question that relates to the student's idea and pushes for something more
Other (Oth)	The follow-up move cannot be analysed by the above codes

particular answer. 'Revoice' reflects the educator's effort to clarify a student's response so that the rest of the class can understand it.

In view of the discourse data collected during the main study phase, it was decided to add the 'confirm', 'press' and 'other' moves, thereby extending the frameworks used in the pilot study data analysis. The foremost move relates to when the educator is not clear about what a student has said and, consequently, checks what she/he has heard, whereas the 'press' moves are those occasions when the educator wants students to clarify, exemplify or justify their explanations. Table 4.5 contains description of the codes applied to the BGEs' follow-up moves.

BGE Questions   DeWitt and Hohenstein's (2010) coding schema for teacher questions was adopted to analyse the questions asked by the BGEs in this study (see Table 4.6). Within their schema, teacher questions are grouped into open-ended and closed-ended questions, with the open-ended questions referring to those that seem to encourage content-related description, reasoning and explanation as well as supporting and scaffolding students' conceptual learning. By contrast, closed-ended questions are those that are factual or procedural in nature and request short answers. In general, as DeWitt and Hohenstein have argued, open-ended questions encourage student initiative and control, whilst closed questions indicate a teacher-centred learning environment and promote less higher-order thinking.

Table 4.6 The coding schema for BGE questions (Adapted from DeWitt & Hohenstein, 2010)

	Questions	Feature	Example
Open-ended questions	Prompt reflection, reason, explanation (PR)	Questions that encourage elaboration of the thinking process, higher-level thinking (i.e. explaining, synthesizing, critiquing, explicating, predicting) or an expansion of ideas	Why is a bloodworm red?
	Prompt description (PD)	Questions that encourage open-ended description based on observations of what is being seen or done, regarding a particular topic	What's the difference between a dragonfly and a damselfly?
	Other open ended (OO)	Questions that are phrased as yes-no questions but may be judged to leave open the opportunity for more reflection than a typical yes-no question or to invite a more in-depth response or reflection	Are you sure?
Closed-ended questions	Right answer (RA)	Questions that call for a short factual answer. Includes questions for which there is more than one possible answer (but there is a fairly limited set of right answers). This code is also used for leading questions, when the BGE has a particular answer in mind	What is the reading on that thermometer?
	Invite participation (IP)	Questions that do not necessarily call for verbal answers but that provide an opening for the student to engage physically with an activity	Would you like to stick the [picture] stem on the wall?
	Check (CK)	Questions that are generally procedural in nature, focusing on the task at hand, or questions that are keeping students on task or making sure they're progressing with the task	Have you done your drawings?
	Clarification (CL)	Questions that ask for clarification or repetition of something that has been said. This category is limited to cases where the questioner is misunderstanding, verifying or checking on understanding	Do you mean…?
	Routine (R)	Questions that are more about the routine than about the subject matter or task	Can you stop chatting please?
	Tag	Used for tag questions	It looks nice, doesn't it?
	Other (OTH)	Questions that do not fit into the above categories, including some seeking a brief 'yes' or 'no' answer. These are often task related but may occasionally relate to content	

Student Questions Borun, Chambers and Cleghorn (1996) investigated family visitors' communication in science museums and classified their learning talk into three: 'identifying', 'describing' and 'interpreting and applying', each of which relates to distinct stages in the process of conceptual learning. Similarly, Fienberg and Leinhardt (2002) grouped visitors' learning talk into 'identification', 'evaluation' and 'expansion', when they explored the conversations of visitors at an exhibition about the history of glass making in Pennsylvania. More specifically, the identification category referred to the visitors' brief statements or questions related to the exhibits, whilst evaluation involved either positive or negative comments made regarding the aesthetic quality of the exhibits, and expansion referred to any extended forms of interaction with an exhibit and this was further divided into: 'analysis', 'synthesis' or 'explanation'. With regard to this, analysis referred to the visitor's explicit or implied comparison of the features of one of the exhibits to something else, whilst synthesis involved the congruence of ideas taken from different exhibit stations or the making of connections between the visitors' previous experiences to an exhibit, and explanation referred to visitors' more comprehensive talk, when they sought to explain how or why something worked or why and how something had happened. From the above frameworks, it appears that the museum visitors' learning talk comprised two categories, the first in which they constructed exhibit-based knowledge, through identification and description, and the second, in which they sought to make connections with their prior knowledge, through: analysis, synthesis and explanation.

In Chap. 3, some of the relevant studies regarding student-generated questions in the science classroom context were reviewed. When information from this particular literature was combined with the evidence presented above regarding museum visitors' learning talk, I was able to propose a coding schema for the identification of the student-generated questions that were recorded during the school trips to the botanic gardens (see Table 4.7).

In the table it can be seen that lower-order questions are those that students ask in order to clarify information, confirm explanations and make exemplification or description. I further coded these questions into 'factual', 'procedural' and 'confirmation' questions, where the foremost require only the recalling of information that identifies or describes something (e.g. What's this?), the procedural questions are used to seek clarification about a given procedure when completing a task (e.g. How many leaves do we need to collect from the ground?) and the lattermost request confirmation of certain information (e.g. Is the plural form of cactus, cacti?).

By contrast, the higher-order questions require answers that go beyond matters of simple information and rather focus on predictions, explanations and problem solving, which can prompt students' epistemic thinking (Ohlsson, 1996; Osborne, 2005). During the process of epistemic thinking, the students can consider and

Table 4.7 The coding scheme for the identification of student-generated questions

Types of student-generated questions	Scardamalia and Beretier (1992)	Watts et al. (1997)	Chin et al. (2002)	Pedrosa de Jesus et al. (2003)
Lower-order questions Factual questions Procedural questions Confirmation questions	Text-based questions	Consolidation questions	Basic information questions	Confirmation questions
Higher-order questions Comprehension questions Prediction questions Synthesis questions Evaluation questions	Knowledge-based questions	Exploration questions Elaboration questions	Wonderment questions	Transformation questions

speculate regarding what they have experienced during the botanic garden visit, by drawing on their school knowledge and daily life experiences. The integration of prior knowledge and new experience reflects students' curiosity and puzzlement and advances their conceptual understanding (Chin & Osborne, 2008). This category was further coded into 'comprehension', 'prediction', 'synthesis' and 'evaluation' questions, according to their functions, with the first involving situations when students seek an explanation of something they do not understand (e.g. Why does a leech suck blood?). The second refers to instances where students speculate or pose a hypothesis (e.g. What would happen if I put my finger into the Venus fly trap?), whilst the third requires students to take two kinds of information and put them together (e.g. How can you distinguish between a dragonfly and a damselfly?) and the last set of questions demand that students set standards of merit and make value judgements (e.g. How can I do a better observational drawing?).

To provide an illustration of how the transcripts were processed for this study, a section of a transcription taken from a published journal paper on secondary science classroom discourse (Scott et al., 2006) has been analysed. Table 4.8 shows an example of when this analytical scheme is applied.

In order to validate the data analysis, another doctoral student was invited to code 15 % of the discourse data collected for the main study by following the same analytical steps as explained above. The Cohen's kappa, which is a statistical measure of inter-rater agreement, was calculated to assess the reliability of the coding scheme, with kappa of the discourse data analysis on average reaching 0.72, which represents a moderate for inter-rater reliability (Gwet, 2001).

Table 4.8 An example of dialogic analysis

		Discourse	Move	Follow-up	Question
1	T:	So, how do you explain it?	I		PR
2	T:	What happens when we feel hot and cold?	I		
3	S2:	Maybe the temperature of the water passes to your hand when you put it in the water.	R		
4	T:	What passes to your hand?	F	MAT	PR
5	S2:	The temperature.	R		
6	T:	The temperature?	F	MAT	PR
7	T:	Do you agree with that?	F	MAT	
8	S5:	There was a heat change.	R		
9	T:	Heat change.	F	REP	
10	T:	What's that?	F	PRS	
11	T:	Can you explain please?	F	PRS	PR
12	S3:	There was a kind of diffusion.	R		
13	S3:	The temperature of the water passes to your hand and from your hand to the water.	R		
14	S6:	One swaps heat with the other.	R		
15	S7:	I think that it's a change of temperature.	R		
16	S6:	The heat warms the cold water until a point at which the temperature will transfer neither cold nor hot.	R		

4.4.2 The Analysis of Pedagogical Behaviours

The data analysis that addresses pedagogical behaviour was taken mainly from the video/audio footage with the field notes providing additional annotation. Through viewing the video and checking the field notes, the participants' behaviours during the class sessions could be added to the discourses that had been transcribed for the earlier dialogic analysis. All the transcripts were uploaded to the QSR Nvivo8 qualitative data analysis software.

The next step of the analysis was to disaggregate all the data into small episodes. For example, an episode could be a BGE's demonstration of using pond-dipping equipment to collect water animals or his/her explanation of the meaning of condensation. When the episodes had been identified, each was repeatedly reviewed until appropriate codes emerged that could be used to code all the episodes that once this iterative process had been completed, some salient themes emerged from the data.

The third step was to design emergent categories to fit all the codes. This process involved merging and expanding the subcategories, that is, for a category that represented the same theme, these were combined so as to redefine the category. Having generated a set of categories from the data, these were discussed with one of my supervisors in order to confirm the validity of the categorization.

In sum, the data analysis in this section followed a bottom-up, grounded theory approach that developed the categories of the BGEs' class management, pedagogical moves and teaching narratives, from the observation data. This approach to data analysis allows the qualitative researcher to integrate the data, research context and latent theories, in an intuitive way (Erlandson, Harris, Skipper, & Allen, 1993). The analysis of the BGE pedagogical behaviours is presented in Chap. 6.

4.4.3 Analysis of the Interview Data

As explained above in relation to the interview protocol, a BGE's answer to a particular question may indicate different aspects of their thinking, for instance, a comment in response to a question about perceptions regarding teaching in the botanic garden context can also include his/her description of teaching strategies. Therefore, the decision was taken to analyse the interviews as a whole, instead of focusing on the individual responses to particular questions.

The analysis of the interview data followed a grounded theory approach (Bodgan & Biklen, 2007; Corbin & Strass, 2008), and in Table 4.9 an example is given of the coding used. By following this coding approach, several common themes emerged from the interview data, and with respect to the BGEs' professional background, the following categories emerged: academic qualifications, teacher training experience, teaching appointments and knowledge of science and botany. With regard to their pedagogical perceptions, three themes were identified: botanic garden as a learning site, hands-on experiential learning and connecting visits to daily life. The BGEs'

Table 4.9 An illustration of the coding for the interview data

	Interview transcript	Coding	Theme
43	*Interviewer:* What do you expect the students		
44	to get from the visit?		
45	*Mark*: We hope they like the things they	enjoyment	
46	learned. We expected them to make sure	being focused	
47	they are listening.		
48	But that's the teacher's job, the discipline.	disciplinary	
49	One of my expectations is that they come	issues	
50	here and they are paying for our services so		Enjoyable
51	they should listen, but that is up to the		learning
52	teacher.	sensory	experience
53	Other expectations were hopefully we'll give	experience	as an
54	them the chance to sense of all about the		objective
55	natural world. So we expected them to be		
56	excited. We fully expected them to be lively		
57	and excited, because it (the garden) is different	novelty factor	
58	from the indoors. They just stepped from the		
59	busy street with a lot of buses and cars going		
60	down and then going into somewhere like this		
61	so we are expecting them to be excited by the		
62	experience of coming through the door.		
63			

responses to the questions about the practical strategies they used to support learning fell into four areas: being enthusiastic, providing sensory experiences, engaging through interactions and collaborating with accompanying adults. More specifically, as I was more interested in exploring the distinctive characteristics of each of the BGE's pedagogical practices, the analysis of the interview data focused on identifying their individual practice.

4.4.4   Analysis of the Impression Sheets

The analysis of the impression sheets also followed a grounded theory approach, through which the qualitative data was coded to elicit themes or patterns. All the sheets collected from the students were scanned into portable document format and analysed with the assistance of Adobe Acrobat®, a software package that can be used to create, manipulate and manage files. The students' writing and drawings were categorized according to the questions listed on the impression sheet. Table 4.10 presents the detailed coding of the survey data.

Table 4.10 The codes developed from the impression sheets

Category	Codes
The most interesting part of the visit	Experiencing various climates
	Direct interaction with fauna and flora
	Attending the BGE workshop
	Hands-on activities
The least interesting part of the visit	Nothing was uninteresting
	Travelling to the garden
	Waiting for activities
	Lunch break
The most favourite part of the BGE teaching	Learning interesting facts
	Seeing different artefacts
	Answering questions
	Explanation
The least favourite part of the BGE teaching	Nothing
	Listening to the introductory talk
	Listening to the BGE speech
	Few opportunities to interact with plants
The number of acts that students remembered	Three
	Two
	One
	None
Students' previous experience in botanic gardens	First time to visit a botanic garden
	More than three times to visit a botanic garden

4.5   Reflections on the Research Design and Methodologies

Patton (2002) contends that in a qualitative research endeavour 'the researcher is the instrument' (p. 14), and throughout this study, I made decisions regarding: what to do and when, what data to collect and what to ignore and what lens to bring to the analysis. It transpired that achieving an understanding of the BGEs' pedagogical practices required researching beyond quantitative measures, and, in fact, it was necessary to accept and appreciate the value of applying interpretive forms of analysis, so as to investigate their professional backgrounds, pedagogical thinking and their communicative approaches when interacting with students. Although a degree of personal bias was inevitable during the research journey, I endeavoured to make my account of the BGEs' pedagogical practices with visiting school groups as real and unbiased as possible. In this regard, I had already formed opinions on the calibre of each BGE from my earlier dealings with them, and so I took great care not to let these judgements have any impact on how I observed any of them.

On the practical side, due to personnel constraints I missed the planning meetings in preparation for the school trips, between the BGEs and schoolteachers, and so the detailed information about those meetings had to be sought during the interviews with the BGEs, which meant it was only second hand. Another limitation was not being able to interview students prior to and after their visits, for I had gained access to schools for this purpose during the pilot study. Before the main study, the BGEs informed me that many schools had refused to participate in the research, because they did not want their students to be filmed and interviewed as well, and so instead of conducting the pre- and post-visit interviews with students, I decided to use the impression sheets to assess their visiting experience.

Turning to the theoretical limitations, as it was discussed earlier, generalizability is the most criticized aspect of the case study method. In this regard, Stake (1994) notes that the case study approach 'emphasizes designing the study to optimize understanding of the case rather than generalization beyond' (p. 236). Therefore, this study is not aimed at developing prescriptions for successful teaching strategies in botanic gardens but rather to provide individual examples of how informal educators can engage and support children's learning during a school trip. My selection of four cases did not allow me to portray the wider variations that occur on other visits and in other botanic gardens.

In qualitative studies, researchers can undertake specific process to ensure trustworthiness, which is an indicator of a study's reliability and validity. According to Lincoln and Guba (1985), trustworthiness in a qualitative enquiry can be 'established by the use of techniques that provide truth value through credibility, applicability through transferability, consistency through dependability and neutrality through confirmability' (in Erlandson et al., 1993, p. 132). In this subsection, the techniques used to establish the trustworthiness of this study are described.

Credibility is concerned with the confidence in the 'truth' of the findings from a particular enquiry (Lincoln & Guba, 1985, p. 290). In this regard, data source triangulation and methodological triangulation were employed in this study to ensure the accuracy of the findings. With respect to the former, Stake (1995) suggests that 'data source triangulation is an effort to see if what we are observing and reporting carries the same meaning when found under different circumstances' (p. 113), and, in particular, for him, data source triangulation enables the researcher 'to see if the phenomenon or case remains the same at other times, in other spaces, or as persons interact differently' (p. 112). Turning to methodological triangulation, this refers to the researcher applying different methods to investigate the same object of study, so as to enhance the validity of the research findings (Cohen et al., 2000). In response to this, multiple data collection methods including interviews, observations, videotaping and a survey were used to investigate the phenomena of interest in this study, and this combination of the different methods provided a basis for checking whether the data interpretations effectively reflected the participants' construction of their realities.

The results of this study may inform other educators and researchers who are interested in teaching ecological science in botanic gardens and even more widely in informal science learning contexts. By following Lincoln and Guba's (1985) recommendation, I implemented two strategies, thick description and purposive sampling, with the purpose of achieving transferability. Thick description may establish an empathetic understanding for the reader, by bringing them vicariously into the context being described (Erlandson et al., 1993; Stake, 1995). The field notes and video recordings enabled me to provide detailed descriptions of the context, participants and phenomena that I observed. With such descriptions, educators in botanic gardens may have 'a sense of déjà vu upon actually visiting the setting' (Erlandson et al., 1993, p. 33), which resonates with their own particular practice. Moreover, my purposive sampling of four BGEs for in-depth case study analysis provides a range of information about pedagogical strategies, some of which could be employed, effectively, both inside and outside of the classroom and for teaching other subject matter than simply ecology.

It is difficult to ensure dependability in a naturalistic study because of the shifts of reality, that is, the participants' viewpoints, behaviours and actions tend to shift under different circumstances. Even if another researcher repeated this study by following the same methodological approach with the same participants, it cannot be guaranteed that they would get the same results. However, during this research journey, the frequent discussions with my supervisors about the methods used to collect and analyse the data and reflections on these helped to form an 'audit trail', which 'made it possible for an external check to be conducted on the processes by which the study was conducted' (Erlandson et al., 1993, p. 34).

In the words of Lincoln and Guba (1985), confirmability is concerned with the degree to which the findings of an enquiry are determined by the focus of the study and not by the biases of the researcher. It is a measure of the objectivity of the research, including what I did, how the research was conducted and how I arrived at the conclusions to the study. With regard to this study, to ensure the research process was transparent enough to be understood by others, I kept detailed records of the data collection, coding and analysis procedures and invited a doctoral student who did not get involved in this study to act as a critical friend, by reviewing and coding some of the data, to check that the processes were being carried out reliably. Furthermore, as a further form of checking, the research processes and some of the preliminary findings were presented to an international science education conference, as this provided the opportunity to receive constructive feedback from a well informed and critical audience.

To sum up, this chapter presents the overall methodology and the specific methods used to investigate the research questions. Initially, the choices and rationale that informed the enquiry paradigm, namely, an interpretivist approach, were discussed. The multi-case study method was decided upon as serving the aims and objectives of this study, although it has limitations such as a lack of generalizability. Next, the difficulties regarding getting access to the research sites and the value of developing a good rapport with the BGEs were outlined. Subsequently, the detailed descriptions regarding the research sites and the participants were provided. Turning to the data, this was collected through observations of the guided school visits, semi-structured interviews with each participating BGE and a student survey using impression sheets. Some of the data analysis was carried in parallel with the phases of its collection, so that the codes could be categorized and linked to the research questions as soon as was possible. Subsequently, once the whole data collection process was finished, these codes were added to and slightly modified. In the last section of this chapter, there has been discussion on the ethical considerations, limitations and trustworthiness of this study.

References

Adler, P. A., & Adler, P. (1994). Observational techniques. In N. K. Denzin & Y. S. Lincoln (Eds.), *Handbook of qualitative research* (pp. 377–392). Thousand Oaks: SAGE.

Bakhtin, M. M. (1986). *Speech genres and other late essays* (V. W. McGee, Trans.). Austin: University of Texas Press

Barron, B. (2000). Problem solving in video-based microworlds: Collaborative and individual outcomes of high achieving sixth-grade students. *Journal of Educational Psychology, 92*(2), 391–398.

Barron, B. (2007). Video as a tool to advance understanding of learning and development in peer, family, and other informal learning contexts. In R. Goldman, R. Pea, B. Barron, & S. J. Derry (Eds.), *Video research in the learning sciences* (pp. 159–188). Mahwah: Lawrence Erlbaum.

Bodgan, R. C., & Biklen, S. K. (2007). *Qualitative research for education: An introduction to theories and methods* (5th ed.). London: Pearson Education.

Borun, M., Chambers, M., & Cleghorn, A. (1996). Families are learning in science museums. *Curator: The Museum Journal, 39*(2), 123–138.

Cazden, C. B. (2001). *Classroom discourse: The language of teaching and learning* (2nd ed.). Portsmouth: Greenwood Press.

Chin, C., & Osborne, J. (2008). Students' questions: A potential resource for teaching and learning science. *Studies in Science Education, 44*(1), 1–39.

Chin, C., Brown, D. E., & Bruce, B. C. (2002). Student-generated questions: A meaningful aspect of learning in science. *International Journal of Science Education, 24*(5), 521–549.

Choi, M. Y. (2007). *Watershed environmental education in South Korea: Understanding learning within communities of practice using social, cognitive and participational theory* (Unpublished doctoral dissertation). King's College London, London, UK.

Cohen, L., Manion, L., & Morrison, K. (2000). *Research methods in education* (5th ed.). London\ New York: Routledge.

Corbin, J., & Strass, A. (2008). *Basics of qualitative research: Techniques and procedures for developing grounded theory* (3rd ed.). Thousand Oaks: SAGE.

Creswell, J. W. (2008). *Research design: Qualitative, quantitative, and mixed methods approaches* (3rd ed.). London: SAGE.

Denscombe, M. (2003). *The good research guide for small scale social research projects* (2nd ed.). Maidenhead, England: Open University Press.

Derry, S. J. (2007). Video research in classroom and teacher learning. In R. Goldman, R. Pea, B. Barron, & S. J. Derry (Eds.), *Video research in the learning sciences* (pp. 305–320). Mahwah: Lawrence Erlbaum.

DeWitt, J., & Hohenstein, J. (2010). School trips and classroom lessons: An investigation into teacher-student talk in two settings. *Journal of Research in Science Teaching, 47*(4), 454–473.

Edwards, D., & Mercer, N. (1987). *Common knowledge: The development of understanding in the classroom*. London: Routledge.

Erlandson, D. A., Harris, E. L., Skipper, B. L., & Allen, S. D. (1993). *Doing naturalistic inquiry: A guide to methods*. Newbury Park: SAGE.

Fienberg, J., & Leinhardt, G. (2002). Looking through the glass: Reflection of identity in conversations at a history museum. In G. Leinhardt, K. Crowley, & K. Knutson (Eds.), *Learning conversations in museums* (pp. 167–211). Mahwah, NJ: Lawrence Erlbaum Associates.

Firestone, W. A. (1993). Alternative arguments for generalizing from data as applied to qualitative research. *Educational Researcher, 22*(4), 16–23.

Gilhooly, K. J., & Logie, R. H. (1980). Age of acquisition, imagery, concreteness, familiarity and ambiguity measures for 1,944 words. *Behaviour Research Methods and Instrumentation, 12*, 395–427.

Gillham, B. (2000). *Case study research methods*. London: Continuum.

Gold, R. L. (1958). Roles in sociological field observations. *Social Forces, 36*(3), 217–223.

Goldman, R. (2007). Video representations and the perspective framework. In R. Goldman, R. Pea, B. Barron, & S. J. Derry (Eds.), *Video research in the learning sciences* (pp. 3–38). Mahwah: Lawrence Erlbaum.

Goldman, R., Pea, R., Barron, B., & Derry, S. J. (Eds.). (2007). *Video research in the learning sciences*. Mahwah: Lawrence Erlbaum Associates.

Gwet, K. (2001). *Handbook of inter-rater reliability*. Gaithersburg: STATAXIS Publishing Company.

Hall, R. (2007). Strategies for video recording: Fast, cheap, and (mostly) in control. In S. J. Derry (Ed.), *Guidelines for video research in education* (pp. 4–14). Chicago: NORC at the University of Chicago.

Hardman, F. (2008). Opening-up classroom discourse: The importance of teacher feedback. In N. Mercer & S. Hodgkinson (Eds.), *Exploring talk in school* (pp. 131–150). London: SAGE.

Jacobs, J., Garnier, H., Gallimore, R., Hollingsworth, H., Givvin, K., Rust, K., et al. (2003). *Third International Mathematics and Science Study 1999 video study technical report* (Vol. 1). Washington, DC: National Centre for Education Statistics.

Jacobs, J., Kawanaka, T., & Stigler, J. W. (1999). Integrating qualitative and quantitative approaches to the analysis of video data on classroom teaching. *International Journal of Educational Research, 31*(8), 717–724.

King, H. (2009). *Supporting natural history enquiry in an informal setting: A study of museum explainer practice* (Unpublished doctoral dissertation). King's College London, London, UK.

LeCompte, M. D., & Schensul, J. J. (1999). *Designing and conducting ethnographic research*. Lanham: AltaMira Press.

Lemke, J. L. (1990). *Talking science: Language, learning and values*. Norwood: Ablex Publishing.

Lincoln, Y. S., & Guba, E. G. (1985). *Naturalistic inquiry*. Beverly Hills: SAGE.

Lobato, J., Clarke, D., & Ellis, A. B. (2005). Initiating and eliciting in teaching: A reformulation of learning. *Journal for Research in Mathematics Education, 36*(2), 101–136.

Marshall, C., & Rossman, G. B. (2006). *Designing qualitative research* (4th ed.). Thousand Oaks: SAGE.

Mehan, H. (1979). *Learning lessons: Social organisation in the classroom.* Cambridge, MA: Harvard University Press.

Miles, M. B., & Huberman, A. M. (1994). *Qualitative data analysis* (2nd ed.). Thousand Oaks: SAGE.

Nystrand, M., Wu, L. L., Gamoran, A., Zeiser, S., & Long, D. A. (2003). Questions in time investigating the structure and dynamics of unfolding classroom discourse. *Discourse Processes, 35*(2), 135–198.

O'Connor, M. C., & Michaels, S. (1996). Shifting participant frameworks: Orchestrating thinking practices in group discussion. In D. Hicks (Ed.), *Discourse, learning and schools* (pp. 63–103). Cambridge: Cambridge University Press.

Ohlsson, S. (1996). Learning to do and learning to understand: A lesson and a challenge for cognitive modelling. In P. Reiman & H. Spada (Eds.), *Learning in humans and machines: Towards an interdisciplinary learning science* (pp. 37–62). Oxford, UK: Elsevier.

Osborne, J. (2005). *The challenge of materials gallery: A discourse based cognitive analysis.* Paper presented at the Annual Conference of the National Association for Research in Science Teaching (NARST), Dallas, Texas.

Osborne, J., & Dillon, J. (2007). Research on learning in informal contexts: Advancing the field? *International Journal of Science Education, 29*(12), 1441–1445.

Osborne, J., Erduran, S., & Simmon, S. (2004). Enhancing the quality of argumentation in school science. *Journal of Research in Science Teaching, 41*(10), 994–1020.

Patton, M. Q. (2002). *Qualitative evaluation and research methods* (3rd ed.). Thousand Oaks: SAGE.

Pedrosa de Jesus, H., Teixeira-Dias, J., & Watts, M. (2003). Questions of chemistry. *International Journal of Science Education, 25*(8), 1015–1034.

Punch, K. F. (2005). *Introduction to social research: Quantitative and qualitative approaches* (2nd ed.). Thousand Oaks: SAGE.

Scardamalia, M., & Bereiter, C. (1992). Text-based and knowledge-based questioning by children. *Cognition and Instruction, 9*(3), 177–199.

Scott, P. H. (1998). Teacher talk and meaning making in science classrooms: A Vygotskian analysis and review. *Studies in Science Education, 32*(1), 45–80.

Scott, P. H., Mortimer, E. F., & Aguiar, O. G. (2006). The tension between authoritative and dialogic discourse: A fundamental characteristic of meaning making interactions in high school science lessons. *Science Education, 90*(4), 605–631.

Shrum, W., Duque, R., & Brown, T. (2005). Digital video as research practice: Methodology for the millennium. *Journal of Research Practice, 1*(1), 1–19.

Sinclair, J. M., & Coulthard, M. (1975). *Towards an analysis of discourse.* London: Oxford University Press.

Stadthagen-Gonzalez, H., & Davis, C. J. (2006). *The Bristol Norms for age of acquisition, imageability, and familiarity.* Bristol: University of Bristol.

Stake, R. E. (1994). Case study. In N. K. Denzin & Y. S. Lincoln (Eds.), *Handbook of qualitative research* (pp. 236–247). Thousand Oaks: SAGE.

Stake, R. E. (1995). *The art of case study research.* Thousand Oaks: SAGE.

Stigler, J. W., Gonzales, P. A., Kawanka, T., Knoll, S., & Serrano, A. (1999). *The TIMSS videotape classroom study: Methods and findings from an exploratory research project on eighth-grade mathematics instruction in Germany, Japan, and the United States.* Washington, DC: National Center for Education Statistics, U.S. Department of Education.

Tran, L. U. (2006). Teaching science in museums: The pedagogy and goals of museum educators. *Science Education, 91*(2), 278–297.

Tunnicliffe, S. D., Lucas, A. M., & Osborne, J. (1997). School visits to zoos and museums: A missed educational opportunity? *International Journal of Science Education, 19*(9), 1039–1056.

Watts, M., Gould, G., & Alsop, S. (1997). Questions of understanding: Categorising pupils' questions in science. *School Science Review, 79*(286), 57–63.

Wellington, J. (2000). *Educational research: Contemporary issues and practical approaches.* London: Continuum.

Wells, G., & Arauz, R. M. (2006). Dialogue in the classroom. *Journal of the Learning Sciences, 15*(3), 379–428.

Yin, R. K. (2003). *Case study research: Design and methods* (3rd ed.). London: SAGE.

Chapter 5
Multiple Perspectives of Botanic Garden Educators' Pedagogical Practices

As Firestone (1993) suggested, 'multi-case studies can use the logic of replication and comparison to strengthen conclusions drawn in single sites and provide evidence for both their broader utility and the conditions under which they hold' (p. 22). Therefore, here four participating BGEs' teaching practices are presented as individual cases guided by two research questions:

(a) What is the structure of a BGE-guided school trip to a botanic garden?
(b) How do the BGEs interact with students in terms of engaging and supporting their learning?

In order to address the above questions, the data collected from observations, interviews and a survey were analysed and four distinct areas for the subsequent investigation emerged: the structure of the guided visit, the BGE's perception of teaching and learning, BGE-student interaction and the students' views about their visiting experience. Next, according to these identified areas, there is explanation of how the raw data were processed and coded into themes so as to be able to address the research questions as well as for subsequent theory development.

For this research, Tran's (2004) framework for analysing the characteristics of museum educator instructed lessons for visiting schoolchildren was adapted, with the content of the guided visits being coded as 'talk', 'demonstration' and 'hands-on activity' segments, based on observation. The primary difference between each segment was distinguished by the form of student participation and engagement. For instance, as described in Table 5.1, 'talk' refers to when students were only required to listen or say something themselves, whereas 'hands-on activity' gave them more freedom to control their learning.

Although categorizing the observed segments in this way provides a detailed description of the content of the guided visits, analysis based on such a framework does not reflect the levels of student engagement. Consequently, Tal and Morag's (2007) schema was adapted to investigate the extent to which the students were involved in learning, with the activities being grouped into 'passive directed', 'active directed' and 'guided exploration'. 'Passive-directed' activities refer to those such

© Springer Science+Business Media Singapore 2015
J. Zhai, *Teaching Science in Out-of-School Settings*,
DOI 10.1007/978-981-287-591-4_5

Table 5.1 Descriptions of the segments of the visit based on observations (adapted from Tran, 2004)

Segment type	Description
Talk	A BGE-led verbal interchange between the BGE and students during a lesson. This was deemed a distinct segment when it involved the BGE and students: asking and answering questions, sharing past experiences or explaining the content related to the demonstration, activity or topic of lesson. There was no physical participation. However, it was not part of the dialogue needed to carry out an activity or demonstration. This meant omission of talk from a segment did not interfere with doing an activity. In other words, giving instructions was not part of talk as directions were necessary for carrying out an activity
Demonstration	A BGE-led whole class activity that involved using objects, role playing and/or student physical participation, with the students joining in individually or as a whole class. Not all students could get the opportunity to physically participate although their attention and verbal input were needed
Student task (hands-on activity)	Student tasks where the BGE assigned the task, gave instructions and materials and decided the length of available time. Groups of two to four students worked together to complete the task whilst the BGE wandered throughout the learning station talking with groups or individual students. The level of need for students to record data varied depending on the task

as a lecture-type presentation, a demonstration and a whole class discussion led by BGE questions, during which students were only allowed to watch and listen to what the BGE did and said. In contrast, 'active-directed' activities involved students more, either behaviourally or intellectually, whereas 'guided exploration' covered those tasks involving collaboration between the BGE and students, with the purpose of the guidance being not to control the students but to facilitate their participation.

As discussed in Chap. 3, the sociocultural perspective of learning highlights the role of discourse in knowledge construction, and the discourse between children and adults is considered as an important lens through which teaching and learning can be assessed in different contexts (DeWitt & Hohenstein, 2010; Edwards & Mercer, 1987; King, 2009; Osborne, Erduran, & Simmon, 2004). In order to investigate the nature of BGE-student interaction during the visit, the transcripts were analysed according to the flow of power in the discourse. In this context, the extent to which the lesson discourse is initiated by students, either sharing personal experience or posing questions for explanation, indicates different levels of students' engagement in the learning process.

Furthermore, the student survey collected on the impression sheet assessed their views on the experience of visiting the botanic garden, and their responses were split into (a) perception of the whole trip and (b) feedback on the session led by the BGE. Firestone (1993) suggests that a case study researcher has the obligation to provide a rich, detailed, thick description of the case so that readers can bridge the gap between the written case and the application setting. Consequently, next each case is presented in turn by following the pro forma described above and subsequently conclusions are drawn regarding the similarities and differences of the participating BGEs' pedagogical practices.

5.1 Case: Mark

Mark was born in Madagascar, but he grew up and received his education in England and at the time of data collection he had already been working in Garden A for 15 years. Prior to joining, he had worked as an outdoor educator in a horticulture institute for 4 years. As an experienced outdoor environmental science educator, he enjoyed a good reputation in the environmental education community as he had been a committee member for several outdoor and environmental education organizations and besides his teaching job he was an environmental photographer. He attributed his passion and enthusiasm for working in the outdoor environment to his deep love of nature.

Despite being a knowledgeable outdoor educator with a BSc in ecology and substantial working experience in horticultural and botanical settings, Mark had never been trained to be a teacher. Instead, he learned how to teach students through observing other educators' teaching practices in similar outdoor settings. However, the lack of formal training experience did not diminish his passion to teach environmental science. When talking about the reason for choosing to have a career as a BGE, he explained that:

> It is a perfect opportunity to teach people about nature and plants generally who may not normally go to natural places and just pass on my enthusiasm to other people. Because people live in the city, they may not often get access to nature. (Interview, May 2009)

He described his mission, as a BGE, as one of encouraging the public to access nature and to get to know more about the environment. Furthermore, he also offered training for the teachers from local primary and secondary schools and had designed several training sessions to help them to use school gardens or the local environment as cross-curricular resources.

5.1.1 Mark's Perceptions of Teaching and Learning Science Outdoors

Mark expressed the belief that certain teaching strategies are helpful for optimizing students' engagement. In this regard, he described his perception of effective teaching and learning in informal settings as 'facilitating hands-on experiences', 'challenging students' thinking', 'promoting peer collaboration and discussion', 'giving choice to students' and 'maintaining relevance'.

5.1.1.1 Facilitating Hands-On Experiences

Research suggests that situated, engaging activities promote intrinsic motivation and self-regulated learning (Paris & Turner, 1994). Learning in informal contexts, such as in botanic gardens, often includes plenty of hands-on activities, which are interesting and curiosity provoking. In Mark's opinion, students can use their multisensory modalities to have direct interaction with authentic biological specimens, such as living plants, animals and plant artefacts:

> We do try to do much of the interactive, hands-on experience for them [the visiting school-children] as possible. By looking at things in the garden, sketching things in the classroom and collecting things around the garden, so they get to make their own sculptures but that really increases their observational skills. So they are using their ears, their noses, their hands, they are using their senses more or less. (Interview, August 2009)

These hands-on activities have a great potential for engaging students' interest in learning about ecological science and sparking their curiosity about the natural world, for through seeing, hearing, smelling and touching, they can easily make sense of their surrounding environment:

> One thing that all kids always love is pond dipping. You know, using nets to dip into the pond and collect things and looking at them. I think that's most children's number one favourite activity. So that gives them chances, even if they can't catch too much, they are still excited to hunt. (Interview, September 2008)

It is notable that Mark's students asked a large number of questions to seek information, knowledge or truth when they were gathering data during the hands-on activities, which was consistent with enquiry-based learning. However, it was observed that he usually offered answers to the students' questions directly, rather than pumping them to develop their own.

5.1.1.2 Challenging Students' Thinking

Apart from the hands-on activities, open-ended, thought-provoking questions can also engage students with the visit. During the interview Mark explained that he preferred to challenge students' thinking through questioning before introducing a fact or explaining a concept:

> I would like to challenge them [the visiting schoolchildren] a bit before introducing a concept to them. For example, looking at the animals in the pond and they are asked about "why does the creature look like see through?" I then ask them "why it might be useful for animals?" or "what purpose does it serve?", or ask them "what might be useful for animals that are almost invisible?" sort of giving the answer straight back up in a different way. That's one thing I do a lot. (Interview, August 2009)

Csikszentmihalyi's (1975) flow theory highlights the fact that learners will not engage with the knowledge acquisition, if the subject matter is too easy, is too hard or bears no relation to prior learning. In this regard, the questions posed by Mark promoted the idea of students connecting their previous knowledge with new information and this was often seen to enhance their intellectual engagement.

5.1.1.3   Promoting Peer Discussion

Mark suggested that working in pairs or small groups, especially when the students were given tasks to complete by themselves, promoted peer discussion and in return stimulated their interest:

> [The students are] also working in small groups because some glasshouses are very crowded. I have to ask them to work in small groups. Group work can be actually quite good because it could get children to discuss things in their own groups. So for example, in the classroom with plants, artefacts, seed pots and all different seeds and fruits for them to look at, we get them into groups of two or four and then they can have discussions in groups. Hopefully [this] will spark off [their] ideas. (Interview, August 2009)

Although he expected students to be become actively engaged when they were discussing with peers, such discussions were not observed as they were beyond the focus of this study. Nevertheless, it was seen that many of the students shared their ideas with each other about what they had seen when they were on the visit. With respect to this, research on teaching and learning in school classrooms has suggested that if student-student interaction is task focused, the students are more likely to be involved in exchanging information rather than discussing ideas (Galton, 1998). However, if group discussion is enquiry based, the learners have more opportunities to develop reasoned arguments (Mercer, Dawes, Wegerif, & Sams, 2004; Osborne et al., 2004).

5.1.1.4   Giving Choice to Students

Mark argued that one benefit of learning in informal contexts is that it supports students' autonomy. He criticized the school education system, which, as he saw it, gives children fewer choices and uses an evaluation method, which is relatively closed ended:

> Possibly stress of our education where special children being tested at very young age, almost like exams for six years old or whatever. I think being able to do things while you can't be wrong, that's quite encouraging. (Interview, August 2009)

Moreover, he stated his belief that guided visits for school groups should involve the students in having substantial choices when they are exploring the botanic garden. In this regard, open-ended, hands-on activities, such as sketching or collecting samples, were offered as being helpful for developing student motivation and self-regulation:

> I think a certain amount of things without right or wrong answers is necessary for learning. I think it's more for their general development. For example, we give them the sticky card to collect things from the ground. One person might find petal flowers interesting or a lot of stones interesting, some people may put more colours on. It brings out something which is very personal to them. ... It's something you can't do wrong because there's no right or wrong answer. I think that's quite important. (Interview, August 2009)

As Paris, Yambor and Packard (1998) suggested, the freedom of making choice 'leads to commitment, deep involvement and strategic thinking with tasks' (p. 269).

According to Mark, giving students the opportunity to decide what to do during sketching and collecting activities can generate their interest in exploring the natural world since 'a choice of goals and effort investment reflects the personal interest of an individual' (Paris et al., 1998, p. 269).

5.1.1.5 Maintaining Relevance

In their study of how school and university students experience and respond to learning activities concerned with environmental issues, Rickinson, Lundholm and Hopwood (2009) propose three lenses, 'emotions and values', 'issues to do with relevance' and 'different viewpoints between teachers and students', as interpretive tools or heuristic devices to look at different aspects of environmental learning. In Mark's case, he put particular emphasis on the importance of relevance in children's learning, seeing it as the key to success:

> [The teaching should provide] Something relevant to the students' life, something that makes the link to their life, so they know about it. So using examples they will be familiar with. For example, starting off a session with a question such as "How many of you have had breakfast today?" and they put their hands up or not put their hands up. Then the second question "How many of you had plants for breakfast?" and again you'll see how many hands. If you can't see many put their hands up, you start then, to show them plants that can be used for breakfast. That makes the relevant link to their life. So, asking questions which directly relate to what they are doing at the moment or what they have done that day. (Interview, August 2009)

Mark's perception of communicating ecological concepts to students is consistent with Barratt and Barratt Hacking's (2008) argument that the environmental curriculum content should be localized so as to make it relevant to children's everyday lives. That is, as other researchers (e.g. Lord & Jones, 2006; Nagel, 1996) have argued, if the focus of learning is on real-life issues, students are more likely to be motivated to achieve deep understanding, rather than just having surface engagement with the learning material.

5.1.2   The Structure of Mark's Guided Visits

The observed school visits guided by Mark mainly focused on the topic of habitats and plant adaptations and he preferred to teach in small groups so he always divided the visiting school group into two subgroups. As explained earlier in this chapter, the segments of the observed visits were coded as 'talk', 'demonstration' and 'hands-on activity', and after comparison of the observed segments, it is apparent that the students spent a larger proportion of time on hands-on activities such as pond dipping, observational drawings and pond life identification than on the two foremost types (see Fig. 5.1).

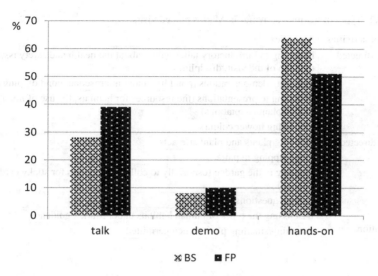

Fig. 5.1 The proportion of observed segments in Mark's guided visits

Furthermore, all the activities on the guided visits were coded into 'passive directed', 'active directed' and 'guided exploration', and Table 5.2 presents the detail of the learning activities that the students from the FP and the BS Schools had according to each category.

The analysis of the time spent on different learning activities (see Fig. 5.2) indicates that Mark's guided visits were child centred, as they offered the students plenty of opportunities to explore either by themselves or with adult guidance. In this regard, the proportion of passive-directed activities for the FP School group was relatively higher than for the BS School group, which may have been due to the fact that the adult to child ratio in the BS School group (1:3) was substantially higher than that for the FP School group (1:10), thus allowing for more free choices, because they could be supported by the accompanying adults.

5.1.3 The Discourse of Mark's Guided Visits

The analysis presented in this subsection investigates the nature of the discourse in Mark's lessons, and Table 5.3 shows his dominant role in this, which left the students very limited space to share their thoughts and thus develop their ideas. In this regard, the table also shows that less than half of Mark's discourse was involved in interaction with the students and hence, overall, his delivery was authoritative and noninteractive in nature.

Table 5.2 Types of learning activities in Mark's guided visits

Types of activities	Activities
Passive directed	Listening to introductory talk (mainly about the health and safety issues, the plan of the visit, disciplines)
	Watching demonstrations (pond life under microscope, pond dipping)
	Listening to presentations (the resources for habitats, the use of medical plants, plant adaptations)
	Recording flower colours
Active directed	Drawing plants and plant artefacts
	Pond dipping in pairs
	Walking in the garden (especially to collect plant parts for sticky card collage)
	Asking questions
Guided exploration	Discussing the types of human habitats under Mark's guidance
	Touching/smelling plants when permitted

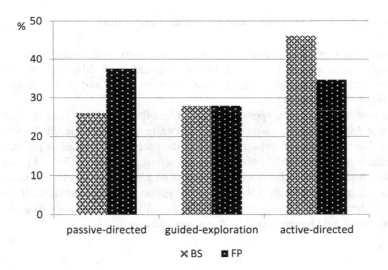

Fig. 5.2 Time spent on different types of learning activities on Mark's guided visits

5.1.3.1   Student-Initiated Discourse on Mark's Guided Visits

Although Mark's talk predominated during the discourse of the observed visits, approximately one third of the BGE-student interaction was initiated by the students (see Table 5.4). In addition, of their utterances, 73 % were classified as volunteering talk, which engaged them in exchanging ideas, posing questions, suggesting

Table 5.3 Mark's (mean) discourse by percentage

School groups	Percentage of Mark's utterances in the lesson discourse[a] (%)	Percentage of Mark's discourse coded as interactive talk[b] (%)
FP	87	36
BS	82	42

[a]The percentage of a BGE's utterances in lesson discourse was calculated as the number of the BGE's utterances divided by the total number of the BGE and the students' utterances
[b]The percentage of a BGE's discourse coded as interactive talk was calculated as the number of BGE's interactive utterances divided by the total number of the BGE's utterances

Table 5.4 Students' discourse by percentage

School groups	Percentage of student-initiated discourse[a] (%)	Percentage of student discourse containing volunteering talk[b] (%)
FP	31	73
BS	33.5	73

[a]The percentage of student-initiated discourse was calculated as the number of student-initiated utterances divided by the total number of the students' utterances
[b]The percentage of student discourse containing volunteering talk was calculated as the number of student volunteering utterances divided by the total number of student-initiated utterances

actions and initiating new topics. Moreover, in spite of Mark's rather authoritative, noninteractive communicative approach, the students did not lose their interest in asking him questions and sharing personal experiences.

The excerpt below is taken from a discourse during the BS School visit, during which Mark was guiding students whilst they were exploring the tropical glasshouse. Here, one student sees a coco de mer tree (*Lodoicea maldivica*) and shares his previous visiting experience at Kew Gardens with his group members.

Transcript			Move
1	S8:	I've seen that before at Kew Gardens! [pointing to a coco de mer tree]	VoL
2	Mark:	Really?	Mnt
3	S8:	The top of the tree against the glasshouse's roof [extending his arm over his head]	
4	Mark:	That's where the double coconuts grow [turns around and talking to another student]	Int

The conversation in the transcript above is initiated by S8, who finds a tree that looks familiar to the one that he has seen before and wants to share his personal experience with others. Mark does engage with encouraging the student's volunteering talk once more by his responding 'Really?', but when the boy can only describe a basic feature about the plant (utterance 3), his follow-up move does not provide any other information to further enhance that student's knowledge about the coco de mer. However, in his defence, it was extremely difficult for Mark to have an in-depth conversation with students on an individual basis given the relatively large size of the visiting group. Another example of student-initiated discourse occurred during the FP School visit, when students were drawing plants displayed on their tables.

Transcript			Move
1	S5:	What's this? [pointing to a plant object on the table]	VoL
2	Mark:	I think that's a dried custard apple	Int
3	S5:	What's the custard apple?	
4	Mark:	It's a type of tropical fruit	Int

In the above exchange, S5 is attracted by an object which has many small bumps on the outside and he seeks help to identify what the plant is, but Mark's answer 'a dried custard apple' (utterance 2) does not offer a full image of the plant nor its purpose. When a further question is asked (utterance 3), he only gives him the very short response 'It's a type of tropical fruit' (utterance 4) and there are no further comments or questions, as the student appears to be satisfied with Mark's answer.

5.1.3.2   Mark's Responses to Students' Talk

As discussed earlier in Chap. 3, the way educators follow up student's responses determines the development of lesson discourse, and regarding this, Table 5.5 presents the different functions of Mark's follow-up utterances. It is notable that the form coded as 'insert' dominated the follow-up discourse, which refers to him responding to the students' request for information directly, rather than pumping them for ideas.

The following discourse from the FP School trip is a good example of the unbalanced power relation in BGE-student interaction, during Mark's guided visits. Here, when he is explaining how succulent plants adapt to the dessert environment, a student notices that there is a cactus covered by white hairs displayed in the corner of the glasshouse and he raised the issue with Mark and the rest of the group.

Table 5.5 Type of Mark's follow-up utterances by percentage ($N_{FP}=389$, $N_{BS}=629$)

School groups	Follow-up moves								
	Mnt (%)	Int (%)	Ect (%)	Evl (%)	Cfm (%)	Rpt (%)	Rev (%)	Prs (%)	Oth (%)
FP	3	75	1	6	2	8	4	0	1
BS	3	74	0	9	4	3	4	0	3

Transcript			Move
1	S2:	Why there are white hairs? [pointing to the old-man cactus]	VoL
2	Mark:	Brilliant!	Evl
3		The one over there in the corner is quite hairy	Rev
4		Actually it would feel quiet soft to us	Int
5		The reason why that plant has got those white hairs is to protect itself from too much light	Int
6		So it's quite useful to be covered by white hairs	Int
7		If it's covered in black hairs that plant would not survive one day in the desert	Int
8		Those of us with dark-coloured hair are probably getting slightly hotter heads than the people with light-coloured hairs [touches blonde and brunette students' heads]	Int
9		Just because dark colours absorb and light colours reflect	Int
10		Just like mirrors	Int

As can be seen the student is curious to see a cactus plant covered by white hairs and asks Mark to explain why this is so (utterance 1). Mark first evaluates the student's question (utterance 2) and rephrases it in the form of a statement to raise the other students' attention (utterances 3–4). Next, rather than throwing a question of his own back to the group, he continues to give a series of further direct statements (utterances 5–10). Admittedly, he does try to provide a simple explanation as to why the old-man cactus (*Cephalocereus senilis*) grows white hair, by comparing the hair of two students' (one blonde, one brunette) to illustrate the relationship between colour and heat absorption. However, by this approach the students' understanding of succulent plants adaptation becomes based on the delivery of factual knowledge, rather than their having to arrive at explanations for themselves through a pressing strategy. There were only a few follow-up utterances that could be coded as 'maintain', which refers to instances that show an educator has attempted extend students engagement by not closing down the exchange during the follow-up moves. The discourse below is a good example, when the students were visiting medical plant beds and Mark told them that the most poisonous plant in the world is called the castor bean (*Ricinus communis*):

Transcript			Move
1	S7:	I know the name of the poison	VoL
2	Mark:	Really?	Mnt
3	S7:	Ricin	
4	Mark:	Ricin	Rpt
5		That's right	Evl

Here S7 volunteers that she knows the name of the poison extracted from the castor bean (utterance 1) and Mark encourages the student to give the answer, by responding 'Really?' (utterance 2). The student then names the poison (utterance 3), which is repeated (utterance 4) and subsequently evaluated (utterance 5) by Mark.

Although the 'maintain' action successfully invited the student to talk more, the follow-up 'evaluative' move brought the conversation to its end.

With respect to the fact that the students' talk was predominantly responded to by direct instruction, in the subsequent interview arranged to reflect on the observed visits, Mark acknowledged that the limited teaching time constrained his practice of cuing and eliciting the students' ideas. Moreover, he expressed his opinion that the students would have gained a more thorough understanding of the subject knowledge if a scaffolding technique could be used to promote their learning dialogue.

5.1.3.3   Mark's Questions

Table 5.6 shows the question types used by Mark during the guided visits, and it should be noted that approximately half of his questions were open ended (prompt reflection, reason, explanation, description), and of the closed questions a high proportion were procedural (check progress, clarification, routine) in nature.

The discourse below gives an example of open-ended and tag questions asked by Mark. When teaching the concept of plant adaptation, Mark begins with the question 'Why you are not wearing your hats, gloves and coats?' (utterance 1), and one student's response, 'Because it's not that cold' (utterance 2), is accepted and confirmed by his tag question 'Because it's not that cold, is it?' (utterance 3). Further information is then provided to support his justification (utterances 4–5).

Transcript			Move
1	Mark:	Why you are not wearing your hats, gloves and coats?	PE
2	S8:	Because it's not that cold	
3	Mark:	Because it's not that cold, is it?	Tag
4		Actually it's about 25 degrees today	Int
5		It's warm enough to be outside in what we are wearing	Int

The purpose of the question is for it to act as a starting point for explaining plant adaptation, by comparing how human beings can change their behaviours to adapt to the surrounding environment. The open-ended question used here (utterance 1) is consistent with Chin's (2007) framing questions, which are used as 'a preface to present small chunks of information which comprises mainly declarative statements' (p. 833).

Table 5.6   Types of questions proposed by Mark ($N_{FP}=39$, $N_{BS}=57$)

School groups	Open ended (%)	Right answer (%)	Procedural and tag (%)	Invite participation (%)	Other
FP	46	10	31	8	5
BS	49	12	35	2	2

5.1.3.4 Students' Questions on Mark's Guided Visits

Table 5.7 shows that most of the questions posed by students were factual and procedural in nature (78 % in the FP School group and 83 % in the BS School group). Despite this predominance of lower-order questions, students did at times ask comprehension, predict, synthesis and evaluate types of questions, which have high-level cognitive demand. The slight variation in the percentage of lower-order questions asked by the two groups of students could be attributed to their age difference, because it is more difficult for younger students to generate higher-order questions than more senior ones.

Among the limited number of higher-order questions posed by the students, the comprehension type provided the greatest proportion. In this regard, according to Bloom, Engelhart, Furst, Hill and Krathwohl's (1956) taxonomy of thinking, comprehension does not require as a high level of cognition as the predict, synthesis and evaluate forms of questions, and hence, this may be the reason for the students asking more of the foremost form than those from the other higher-order question categories.

5.1.4 Students' Views of Mark's Guided Visits

On the impression sheets completed by the Year 5 students from the FP School (39 copies were given out and 36 were returned, thus giving a response rate of 92 %), 27 students (75 %) responded that the trip was their first time to a botanic garden. For the students who had been to Garden A several times, they reported that they had a lot of fun during every single visit.

When considering their answers to the question 'What was the most interesting part of the visit?' 15 replied that this was seeing exotic plants, especially tropical plants and carnivorous plants, which they would not be able to see in school or at home. For example, two responses were as follows:

> The most interesting part of the trip was the tropical greenhouse. There was interesting things growing like banana, vanilla, tamarind, all sorts of things from another part of the world.
> The most interesting part was learning about the foreign plants because some of them were carnivorous.
> Another eleven students wrote that the hands-on activities, such as observational drawing, pond dipping and collecting things, were their favourite sections of the visit:

Table 5.7 Types of questions generated by students ($N_{FP} = 85$, $N_{BS} = 83$)

School groups	Lower order (%)	Comprehension (%)	Predict (%)	Synthesis (%)	Evaluate (%)
FP	78	14	6	1	1
BS	83	10	2	4	1

The most interesting part was looking at the plants and drawing inside the classroom, because it was fun drawing the wasps nest, the sugarcane and the birds' nests.
The most interesting part was pond dipping, because we had a leech, which was friendly so we could touch it.
The most interesting part of the garden visit was sticking the petals and making a shape. I liked it because it was creative and fun.

There were a few other students who stated that the sensory interaction with plants and the discovery of some amazing facts were their most favourite part of the visit. For example:

I think the most interesting thing in the garden was all the plants and all the flowers, because I liked smelling them.
I found looking at all of the chilies was interesting, because I did not know that they could grow so big.

With regard to the responses to the question 'What was the least interesting part of the visit?' more than half of the students stated that they enjoyed the entire visit. However, several students expressed the view that staying inside the classroom, especially listening to Mark's instruction and doing observational drawings, was not very interesting. In addition, others were of the opinion that the recording of flower colours by ticking boxes was the least interesting activity. Thus, it would appear that the students' views of the visit were largely determined by the extent to which they were engaged with the hands-on, exploratory activities.

In addition, many of the students viewed Mark as a kind and friendly educator, who was always sharing exotic and rare things as well as explaining about them with clear descriptions. Moreover, his explanations of ecological science were evaluated very positively by the students, with several responding that his talk about carnivorous and medicinal plants was interesting and easy to understand. Further, the way in which Mark answered their questions was also highly appreciated, for example:

I liked the way he answered all the questions you asked him and he went into them with great description.

Responses like this can probably explain why the students posed many questions on the visit. Apart from the detailed explanations, the students were very pleased that Mark gave them the opportunity to have sensory interactions with many exotic and rare objects that they could not get access to during their normal daily lives, as implied by the following:

I liked the lesson because you could touch all the different nests.
I liked the bit when he gave us some seed pods. It was interesting to see all the different types of stuff.

Although there was much positive feedback, a few students made some critical comments about Mark's teaching, which covered two aspects. The first was regarding the management of the lesson:

Sometimes he started talking before we were all there.

What the student is complaining about here is that Mark began his teaching before assembling all the students together, after they had walked from one learning station to another. On the positive side, such a complaint reveals that this particular student valued the spoken explanations, thus implying that they were strongly engaged with the content of the lesson. The other negative comment relates to the content of the activities and in this case one student wrote this on the impression sheet:

I thought he could have done a shorter tour, maybe doing some activities on the way

It is not clear why this student thought the length of the tour was long and what kind of activities should be introduced. However, from the observations during the tour, especially in the glasshouses and at the medical plant beds, the activities were not highly child centred, with Mark's explanations and walking between objects of interest predominating during most parts of this particular section of the visit. Other students expressed the view that they had few opportunities to touch the plants:

I didn't like how we were not allowed to touch the plants because I really wanted to feel what they felt like.

Regarding these matters, Mark explained during his interview that health and safety issues were a major concern and this was why the students were not allowed to explore the garden by themselves or to touch plants, which they might damage or be harmed by.

5.1.5 Case Summary

It was observed during the guided visits that Mark fulfilled his teaching purpose stated in the interviews of encouraging students to have direct experience in a natural environment and for him to demonstrate to them the many uses of plants. More specifically, he provided them with plenty of opportunities to use their sensory modalities to interact with nature, for example, smelling eucalyptus leaves, touching lamb's ears (*Stachys byzantina*) and observing flowers of different colours. In addition, the discussions about plants eaten for breakfast and the walks in the garden to look at medical plants, vegetable plants and perfumery herbs helped the students to make sense of the various uses of plants.

The analysis of the interaction between Mark and the children has revealed that the lesson discourse was dominated by his authoritative, noninteractive talking style. Moreover, although approximately half of his questions were coded as open ended, his strategy for following up students' responses failed to direct the lesson discourse towards an ongoing dialogue. However, the students' willingness to converse was not discouraged by his way of teaching, as evidenced by the high proportion of them who volunteered to join in and this provides clear proof of their strong level of engagement and intense interest in learning about plants.

5.2   Case: Simon

Simon was employed by the local city council and responsible for the outdoor learning centre based in the Garden B. Of the visiting school groups, Simon only taught those from urban areas, whilst the others were taught by Debbie, who was employed by Garden B's Education Department. However, although Simon and Debbie were employed by different organizations and had different remits, they shared an office and discussed teaching issues every day.

Simon had a BSc degree in applied science, for which he specialized in chemistry and environmental science and after completing it he had taken a PGCE course in secondary school science teaching. Rather than teaching secondary students, he had worked as a primary teacher in several urban schools for 15 years and subsequently became an outdoor educator at a residential centre for 3 years before he moved on to work at Garden B in 2002. He explained the reason for choosing to be a BGE as:

> The reason why I like working here rather than working in schools is you don't see the same children all the time. It keeps you fresh in your head what teaching and learning is about. Whilst in school sometimes you feel very guilty delivering exactly the same thing. (Interview, July 2009)

As an outdoor educator employed by the local education authority, he had plenty of opportunities to receive professional development training so as to keep him updated on National Curriculum innovations and other relevant new initiatives. Moreover, the networking of outdoor learning centres by the city council also gave him chances to reflect on his teaching practice, in particular, through observing other outdoor educators' lessons.

5.2.1   Simon's Perceptions of Teaching and Learning Science Outdoors

As an experienced educator, Simon had developed his unique perception of teaching and learning from 25 years of pedagogical practice in different contexts, and from the interviews with him, five themes emerged: 'learning with fun', 'experiencing another learning style', 'making mistakes, 'interaction as a key to engage children' and 'developing the language of science'.

5.2.1.1   Learning with Fun

Enjoyment is a primary expectation that Simon liked the students to achieve on their visits. Moreover, he expressed the belief that talking to students with a sense of humour is also an important strategy to engage them in listening and when he was teaching he made a lot of jokes to make the visit more interesting. For example, when introducing the concept of photosynthesis, he first compared human beings who need food to survive with plants that can make their own. More specifically, rather than directly explaining what photosynthesis is, he invited a student to hold a dried bamboo stem and asked the class 'If I put this child outside and planted her, two weeks later what would happen to her?' After several turns of prompting, one student answered that 'she would die because there is no food' and then Simon introduced the fact that plants use photosynthesis into the lesson. He explained that using such analogies to present science concepts came from his personal learning experience:

> When I am trying to learn things I also have to associate things often with a picture in my own mind. Different people obviously learn in different ways. But having that association I think it's really useful, something other than just the written word, because a lot of people don't respond to that. (Interview, July 2009)

5.2.1.2   Experiencing Another Learning Style

As Simon contended, one advantage of visiting an informal setting, such as a botanic garden, is that students can get themselves involved through using their eyes and hands to experience things that they cannot in school. Moreover, he explained that a garden is ideal, because the learning experience can cover people who have different learning styles:

> Some people are very physical, so they are a sort of bodily-kinesthetic learners; whereas some people are very visual, so they need things to be written down. (Interview, July 2009)

He criticized the current school education system, which in his opinion has failed to include every child, especially those with learning difficulties and argued that learning that takes place in informal settings is more socially inclusive.

> One of the biggest things about visit centers like this is to give them some spaces to think whereas often in schools they are told what to think. That's why you often get children who perform badly in class but do well on visiting here. (Interview, July 2009)

From this, it can be seen that Simon believed strongly that his form of education delivery transcended what he saw as the unfair setting of the classroom, by providing opportunities for less academically able students to become engaged in learning that did not simply involve written output.

5.2.1.3   Making Mistakes

When responding to the questions regarding what education should BGEs provide to the visiting school students, Simon argued that ideally it should allow children to make mistakes in the process of learning and development, stating:

> They need the chance to get it wrong. They need to push the boundaries and go beyond where they are. That means to make mistakes. That's not always a bad thing always. (Interview, April 2009)

Simon argued that allowing children to make mistakes in their learning should be seen as positive, because:

> They should be brave enough to do that rather than just sit down and stay on the medium level. That means they have to understand they made mistakes. They've got to offer something in order to be challenged otherwise it's wasting their time. (Interview, April 2009)

It was observed that in his practice Simon supported students' thinking by constant eliciting and prompting. For instance, the excerpt below describes how he used this approach to get Year 3 students to grasp an understanding of the types of habitat that succulent plants live in.

Transcript			Move
1	Simon:	Does anybody know what sort of habitat in the world do these plants come from? [points to the cacti]	RA
2	S2:	Ahmmm, Mexico	
3	Simon:	It's a good idea	Evl
4		But it's a sort of habitat	Ect
5	S7:	Is it the errr the…	
6	Simon:	The soil is stony and sandy	Ect
7		Big clue, isn't it?	Tag
8	S13	Desert?	
9	Simon:	Desert	Rpt
10		They all come from deserts in the world	Int

The above excerpt shows that the students are being given opportunities to make mistakes (utterances 2 and 5). Moreover, rather than telling them the answer, Simon prompts them by giving different clues (utterances 3 and 6) until the correct answer is provided (utterance 8). The strategy adopted here is consistent with Edwards and Mercer's (1987) 'cued elicitation', whereby the teacher seeks information from students by providing visual clues or verbal hints. By doing so, Simon is avoiding the mere provision of the missing information or giving a lecture on what kind of environment cactus plants live in and is placing the emphasis on the children doing the learning for themselves.

5.2.1.4 Engaging Children Through Questioning

Viewing learning as an active process rather than a passive one, as shown above, Simon was also of the opinion that during interactions with groups of children, it was critical to get everyone involved and for him the most effective way to engage students in exchanging ideas during guided visits was through questioning:

> The way I do that [engage students] is to continuously question them. So you usually get the responses, but you can't assume that they receive it, but checking, checking and checking [through questions]. (Interview, April 2009)

When reflecting on the questions asked during the observed lessons, he explained that he liked to use open-ended ones as much as possible, because this encouraged different points of view from the students:

> To me they are open-ended questions because I accept any reasonable questions. The point for me to question children is to find out where they are conceptually about whatever they are talking about. So give them a chance to do that. (Interview, July 2009)

Further, he pertained that a successful guided visit should not only involve questions from him but also ones from the students. However, it was observed that there were few student-initiated questions and when asked why he thought this was the case, Simon explained that:

> I was trying to encourage them to ask their own questions. The curriculum in some schools can be very narrow and the kids are not encouraged to ask questions. I think they need to be trained to do that at a very early age. (Interview, July 2009)

In addition on this matter, Simon said that he found it difficult encouraging students to ask their own questions during this particular visit, because of the large size of the group and his unfamiliarity with the students on an individual basis:

> It's not easy for a group of 30 kids. You don't know their personality. You know about them hopefully from their teacher but you don't know them individually. (Interview, July 2009)

Simon was the only participating BGE who used worksheets during the observed lessons, which he explained was because this was requested by the schoolteachers. He, however, stated that he would prefer not to have to do so, because in his opinion, it disengaged students, and so in order to minimize the negative impact brought about by having to use them, his professed strategy was to encourage the schoolteachers to participate actively in their students' exploratory work and discussing with them the possible responses to the questions on the worksheet. In this regard, in their research on school trips to science museums, DeWitt and Osborne (2007) suggest that museum educators should encourage joint productive activity, which 'involves students working with each other and with the teacher towards an end product' (p. 690). Moreover, if the students are engaged in discussing with their schoolteachers, the visit would be more learning oriented than task oriented (Griffin & Symington, 1997).

5.2.1.5   Developing the Language of Science

Apart from promoting classroom interaction, Simon emphasized his interest in developing children's language ability even though there was no such requirement from the schoolteachers when they scheduled the visit:

> Sometimes schools actually say to me like this "we're going to do this", but what we are really interested in is developing the language of children or developing their communication between themselves or cooperation where they would not normally get the chance to do that. (Interview, April 2009)

The Garden B is located in a city centre, which has a large Pakistani and Bengali immigrant community and hence, most of the children from the local schools speak English as an additional language. One of the observed visiting groups (SB School group) was from this area, with several languages being spoken by the students. This language factor remained a big challenge to Simon in his teaching even though he had been working in urban schools for 15 years. In this regard, he explained that sometimes he had to introduce completely new sets of words to the students, whose first language was poor in its coverage of different objects:

> Take a word like "soil" for example. In Punjabi there is one word which is "mitti" which means sand, soil, dirt, dust, anything like that is all one word, just "mitti". So if there is no such word in your own language, you don't actually know the difference between such things. I have to give them extra information which seems like overload. (Interview, July 2009)

According to Simon, English as an additional language was not the only barrier to explaining science to young children, because in his opinion school education seldom developed students' scientific thinking and reasoning, but rather focused on learning by repetition:

> They knew a lot of vocabulary, but they have very little understanding about the process at all. I am aware of that from teaching in mainstream schools. The problem is that we don't measure their understanding. We just tick the box particular for the SAT[1]s. There's no process involved at all. They just learn the words. (Interview, July 2009)

The promotion of science literacy places emphasis on talking science, which is different from the rote memorization of science terms (Wellington & Osborne, 2001). Talking science is not the simple use of scientific words to describe objects and phenomena, but the practical applications of science language to 'express relationships between the meanings of different concepts' (Lemke, 1990, p. ix). Therefore, science educators have the responsibility to develop children's language of science through scientific reasoning and argumentations (Millar & Osborne, 1998).

5.2.2   The Structure of Simon's Guided Visits

The observed guided visits were designed to support the children's science learning based on the QCA's science scheme: 'Unit 3B: helping plants grow well'.[2] According to Simon, the purpose of his teaching was to offer visiting school groups botanical experiences and to reinforce understanding of some basic science concepts, such as 'plants for food', 'plant parts', 'water and plants', 'plants and light', 'plants and

---

[1] National Curriculum assessments are a series of educational assessments, colloquially known SATs, used to assess the attainment of children attending maintained schools in England.

[2] 'Unit 3B: Helping pants grow well' is one part of QCA science scheme for Key Stage 1 and 2 students. However, with the change of primary National Curriculum and the British government in 2010, the schemes on QCA Standards Site do not reflect current government policy.

light' and 'plants and warmth'. As is shown in Fig. 5.3, the segments coded as 'hands-on activity' were most frequently observed, but these activities were highly structured and lacked free choice. For example, Simon only gave the students 5 min to find and draw each plant requested on their worksheet and consequently, they were in too much of a rush to be able to spend time making careful observations.

Table 5.8 shows the learning activities during Simon's guided visits, coded previously as 'passive directed', 'guided exploration' and 'active directed', and the activities based on the worksheet could be considered as being an integration of all three forms. That is, it included both open-ended and closed questions that the students had to provide clear answers for, such as recording the temperature and the level of humidity in the glasshouses, which can be classed as passive-directed activities. Moreover, there were guided explorations when the students were given the freedom to walk around the glasshouse and sketch the plants, during which they were assisted by Simon and the accompanying schoolteachers. Finally, when the children were asked to identify plants with particular clues and subsequently make collages of them, this was an active-directed task.

Figure 5.4 shows the percentage of time that Simon spent on organizing different learning activities. It is notable that the SB School group spent a greater proportion of time on passive-directed activities than the NP School group. This might be because a large number of students from the SB School group did not speak English as their native language and so Simon had to explain the instructions on the worksheet in more detail. Further, the adult to student ratio in the NP School group was relatively higher, so it was possible for the students to receive immediate assistance from the adults. Consequently, the students had more power to control their learning rather than simply passively follow the instructions given by Simon.

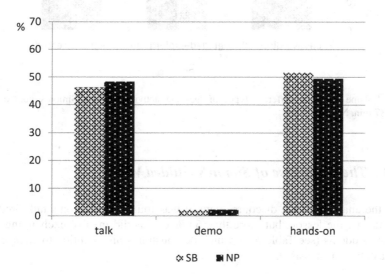

Fig. 5.3 The proportion of observed segments in Simon's guided visits ($N_{SB} = 89$ min, $N_{NP} = 93$ min)

Table 5.8 Types of learning activities on Simon's guided visits

Types of activities	Activities
Passive directed	Listening to the class management instructions (especially about the discipline)
	Listening to explanations of how to use worksheets
	Recording the readings on the thermometer and the rainfall gauge on the worksheets
Active directed	Drawing whilst observing
	Asking questions
Guided exploration	Reviewing previous knowledge about plants and building up the picture of the flowering plant
	Touching/smelling plants when permitted
	Exploring the glasshouse under guidance

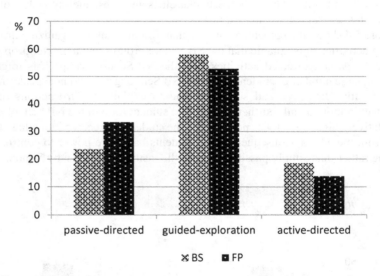

Fig. 5.4 Time spent on different types of learning activities during Simon's guided visits ($N_{SB} = 87$ min, $N_{NP} = 81$ min)

5.2.3   The Discourse of Simon's Guided Visits

From the analysis of the discourse data, it was found that Simon's talk predominated during the lesson, but more than 54 % of this did involve exchanging ideas with the students (see Table 5.9), which demonstrates his intention to engage with them as much as possible.

Table 5.9 Simon's discourse by percentage

School groups	Percentage of Simon's utterances in the lesson discourse (%)	Percentage of Simon's discourse coded as interactive talk (%)
SB	84	54
NP	84	55

Table 5.10 Students' discourse by percentage

School groups	Percentage of student-initiated discourse (%)	Percentage of student discourse containing volunteering talk (%)
SB	9	12
NP	16	23

5.2.3.1   Student-Initiated Discourse on Simon's Guided Visits

Although the discourse in Simon's guided visits was interactive, student-initiated discourse was relatively low and the proportion of that containing volunteering talk was small as well (see Table 5.10). Perhaps, this result might relate to the students' background. The students in the NP School group contributed more to the discussion, which could be due to the fact that all of them were English native speakers, thus showing that the language barrier has a negative impact on the students' participation in verbal communication aimed at constructing shared knowledge.

Even though only a limited amount of student volunteering talk was observed, some of these rare examples are presented next to illustrate what forms these took. In the following discourse, a student is commenting on the feature of lamb's ears (*Stachys byzantina*) when Simon checks his observational drawing.

Transcript			Move
1	Simon:	Have you finished your drawing?	CK
2	Ss:	Yeah	
3	S11:	Some of the leaves are very soft	
4	Simon:	Some of them are very soft	Rpt
5	Simon:	If you imagine you were a fly, would you like to walk over a leaf with hairs on?	Int
6	Simon:	You wouldn't, would you?	Tag
7		Your legs will get caught	Int
8		So it protects themselves from flies walking all the time	Int
9		So it's a trick of nature	Int

After the student's comment on the leaves based on his direct sensory experience (utterance 3), Simon repeats S11's comment and goes on to explain that the lamb's ears needs its soft leaves to protect it from insects. Simon then attempts to interact with the student; however, his tag questions (utterance 6) fail to encourage the student to participate in the conversation. In the next subsection, Simon's approach to follow up student's talk is further discussed.

Table 5.11 Percentage of Simon's follow-up utterances ($N_{SB}=313$, $N_{NP}=272$)

School groups	Follow-up moves								
	Mnt (%)	Int (%)	Ect (%)	Evl (%)	Cfm (%)	Rpt (%)	Rev (%)	Prs (%)	Oth (%)
SB	8	36	8	17	0	17	11	2	1
NP	10	44	9	12	2	15	7	0	1

5.2.3.2   Simon's Responses to Students' Talk

The strategy to follow up student's talk that Simon employed in the above discourse involved a direct explanation, and this was what he most frequently used in his teaching practice (see Table 5.11). Moreover, he often resorted to the 'evaluate' and 'repeat' moves, which have been proved by classroom researchers to be ineffective in developing lesson discourse towards a dialogic form (e.g. Mercer & Littleton, 2007; Scott, Mortimer, & Aguiar, 2006).

By contrast, the 'elicit' and 'press' moves, which have the potential to promote students' reasoning and critical thinking, were infrequently used during Simon's lessons. The following discourse took place when Simon is checking students' prior knowledge of flower structure. After several students have put root, stem, leaf and petal shapes of cards onto a board to build up a flowing plant, S2 notices that something is missing on the picture.

Transcript			Move
1	S2:	You forgot the middle thing	VoL
2	Simon:	I forgot the middle thing	Rpt
3		This young lady thinks we missed the middle thing	Rpt
4		This middle thing on this picture is yellow	Ect
5		This middle thing would have a special powder	Ect
6		Do you know the name of this special powder?	Prs
7		Pollen	
8		It's pollen	Rpt
9		Well done	Evl

With S2's comment, Simon does not give direct corrective feedback, but instead raises the other students' attention (utterances 2–3) and attempts to elicit from the group what the missing middle thing on the picture should be with some clues (utterances 4–5). He then further challenges them by asking 'Do you know the name of this special powder?' (utterance 6). Finally, another student answers the question correctly (utterance 8) and thus, the follow-up strategies that Simon employed in this exchange did encourage the students to develop their thoughts rather than receiving the answer passively.

It should be noted that approximately half of the student-initiated talk was related to the identification of plants and that Simon did not always give direct feedback to those questions. Instead, he often threw the responsibility of thinking back to the students and encouraged them to find answers by themselves, for example:

Transcript			Move
1	Simon:	Children, shall we meet at the dragon tree that I asked you to find out earlier?	
2	S8:	Where is it?	
3	Simon:	Which one do you think is the dragon tree?	Mtn
4		I'll give you a clue: the only proper tree here	Ect

Simon requests that the students to get assembled at the dragon tree as most of them have finished the worksheet (utterance 1); however, one student is confused and asks where the dragon tree (*Dracaena*) is (utterance 2). Rather than telling her directly, Simon provides the girl with a clue to help her identify the object (utterances 3–5).

5.2.3.3 Simon's Questions

As highlighted earlier in this chapter, Simon considered questioning to be an important strategy to engage children in thinking and thus support learning. In practice, he did pose a lot of questions, but most of them were closed ended (see Table 5.12), with the 'procedural', 'tag' and 'right answer' questions representing more than half.

The following excerpt displays an example of Simon reviewing the children's previous knowledge by posing right answer questions. Before the glasshouse exploration, Simon was reviewing the students' knowledge of plant structure and the use of the scientific term for each part of the plant was emphasized highly in his talk. However, some students were still using basic language to describe plants when they got to the glasshouse.

Transcript			Move
1	Simon:	What part of the plant did I touch?	RA
2	S8:	The green bit	
3	Simon:	The green bit	Rpt
4		We've studied that in the classroom	Ect
5		The part that I'm touching is called…? [pauses for students' response]	RA
6	Ss:	Stem	
7	Simon:	This is the stem	Cfm
8		That's right	Evl
9		Remember you use those words	

The question 'What part of the plant did I touch?' (utterance 1) sounds like a procedural question, but the purpose is to find out whether the students can express the terms they have learned in a practical setting. The question was coded as a 'right

Table 5.12   Types of questions put by Simon (N_{SB} = 123, N_{NP} = 93)

School groups	Open ended (%)	Right answer (%)	Procedural and tag (%)	Invite participation (%)	Other (%)
SB	27	26	34	8	5
NP	38	15	38	7	2

Table 5.13   Types of questions generated by students (N_{SB} = 8, N_{NP} = 15)

School groups	Lower order (%)	Comprehension (%)	Predict (%)	Synthesis (%)	Evaluate (%)
SB	87.5	12.5	0	0	0
NP	93.3	0	6.7	0	0

answer' question, because there was only one acceptable answer in Simon's mind and the student's response 'the green bit' was not acceptable, especially when they had learned its scientific term, 'stem'. Simon further responds 'The part that I'm touching is called' in a questioning voice and pauses for the students' response. What Simon expected here was that the students could fill in the blank to complete the sentence, which is similar to the 'verbal jigsaw' as defined by Chin (2007), used 'to elicit the appropriate and essential words from students for the construction of declarative knowledge in the form of a network of related concepts' (p. 826). On the observed guided visits, verbal jigsaws were widely used to emphasize keywords and phrases, especially with the SB School group who were not very articulate or verbally expressive.

The excerpt discussed above also contained some right answer questions, for instance, 'Which one do you think is the dragon tree?' and 'Do you know the name of this special powder?' and despite of their not being open ended, they did successfully prompt the students' thinking and encouraged them to talk. These questions are similar to the 'reflective toss' questions, described by van Zee and Minstrell (1997), which, unlike triadic dialogue controlled by the educator, invite students into conversation by building on an initial utterances from them. Chin (2007) characterized this form of questioning as Socratic questioning, which can 'help teachers shift toward more reflective discourse that helps students to clarify their meanings, consider various points of view and monitor their own thinking' (p. 818).

5.2.3.4   Students' Questions on Simon's Guided Visits

Compared to the numbers of questions asked by Simon, student-generated questions, coded as 'higher-order' questions, were extremely rarely observed (see Table 5.13). When reflecting on his teaching during the second interview, he complained that the rote teaching commonly found in school education does not support student questioning and as the students were not used to asking questions in classrooms, they were not likely to pose higher-order questions during informal school trips.

5.2.4 Case Summary

As pointed out in Sect. 5.2.1.5, Simon's greatest interest in teaching visiting school groups was to facilitate children in acquiring the language of science. In this regard, Wellington and Osborne (2001) argue that language is a major barrier to most students when learning science, because 'one of the important features of science is the richness of the words and terms it uses' (p. 3). On the observed guided visits, much of teaching focused on the use of these terms in practice. For example, Simon asked a lot of 'verbal jigsaw' questions to force students to use science terms to describe plants. The findings from this case study also suggest that the students who speak English as an additional language have difficulty in participating in verbal communications, but the 'elicit', 'maintain' and 'press' feedback moves may encourage them to break through the language barrier through the dialogic interactions with the educator that these can generate.

5.3 Case: Debbie

As introduced earlier, Debbie was a full-time BGE employed by the Garden B, who was responsible for the school groups outside of the city. She had never worked in mainstream schools, although she had been trained and qualified as a secondary teacher. Before she started her profession as a BGE, she had been working as an outdoor educator for 6 years in different environmental education organizations, including for the Royal Society for the Protection of Birds (RSBP), Royal Society for the Prevention of Cruelty to Animals (RSPCA) and Wildlife Trusts. At the time of the data collection, she had been working in Garden B for 5 years; however, she still did not have much confidence in teaching ecological science:

> I'm always looking to improve my plant knowledge. I still feel I have a lot to know. (Interview, June 2008)

She had found that working at Garden B was very helpful for her professional development, in particular because it gave her the opportunity to further develop her content knowledge. She also stated that she had been to the Botanic Garden Education Network (BGEN) conferences for 4 years, where she networked with other BGEs so that they could learn from each other. Moreover, she was a member of the Group for Education in Museums (GEM) and had attended the training for facilitating public engagement.

According to the observations, her guided visits usually lasted 45 min and then the rest of the visit was either led by education volunteers or taught by accompanying schoolteachers. As she explained, the school education programmes were designed to reinforce students' prior knowledge of plants as well as to stimulate them to be focused during the visit, even without the guidance from garden education staff and accompanying schoolteachers.

5.3.1 Debbie's Perceptions of Teaching and Learning Science Outdoors

Debbie's education beliefs were greatly shaped by her previous working experience, and her perception of teaching and learning in outdoor settings had four themes, including 'contextualized teaching', 'motivating students through interactions', 'the National Curriculum as a guide rather than constraint' and the 'time factor as a teaching barrier'.

5.3.1.1 Contextualized Teaching

The guided visits observed for the main study were focused on the 'Amazon Rainforest' and despite the fact that Debbie had taught this topic to many groups of students, she never took it for granted that she would just repeat her lesson each time:

> You can't stand in front of the class and read the script. You can't have a script in your head, as such. (Interview, July 2009)

Instead she preferred to modify her teaching according to the context of each school group and the special requirements asked of her by the schoolteachers. From her point of view, the most challenging occasions were when the students' level of knowledge fell behind what their schoolteachers had reported prior to the visit. Consequently, she often had to modify the content of the lessons and deliver them with appropriate teaching approaches on the spur of the moment:

> You have to continuously adapt because you talk to schoolteachers before they come. They said "yes we've done rainforests". The group arrives but show you, actually whatever they've done at school, they haven't remembered for some reason. (Interview, July 2009)

5.3.1.2 Motivating Students Through Interactions

To Debbie, students' engagement and active participation were important, if there were to be a productive learning experience. In this regard, she pointed out that she did not want students to sit and listen to her all the time but for them to get involved as much as possible. Moreover, she believed it was essential to make every student feel part of the lesson:

> If you are going to do something and you don't respond to the audience, you don't take into the account the response you should get and you just do what you are going to do. Anyway, it's not going to work. (Interview, July 2009)

Further, she explained that she aimed to engage students by inviting them to come to the front of the class and to assist her during demonstrations. She had also found that making jokes was an effective way of connecting with students, as the following shows:

> You know we get the biggest seed in the world, the double coconut. It looks like a bottom. So we say "can we say the word 'bum'?" We make a joke but they will remember the fact that the biggest seed in the world looks like a bottom. (Interview, July 2009)

Through such social interactions, she expected to make the teaching session not only more fun but also memorable and worthwhile. Apart from getting students involved, she believed that making the schoolteacher a member of the workshop was helpful for her teaching, as their participation helped the class to recognize the importance of their needing to learn and so they concentrated better.

5.3.1.3   The National Curriculum as a Guide

Debbie explained that the education programmes for visiting school groups were developed under the guidance of the National Curriculum. However, she added that the design and delivery of the visits were not only based on the National Curriculum, but also had to take into account the schoolteachers' requirements and the facilities that the garden had.

Debbie preferred to offer schoolteachers and students free choice; thus, she tended not to give detailed directions and instructions about the activities that they could do on the visit:

> We never tell the school what they have to do when they are here. They are very much in charge of their own learning and their own aims and objectives. (Interview, July 2009)

Sometimes the schoolteachers did not express any specific objectives, especially during summertime when the students were free from SAT preparation, and under such circumstances, the schoolteachers often considered the school trip as just a fun day out with minimum learning expectations. However, whatever were their intentions the schoolteachers were always provided with the education packs prior to the visit, which gave the range of activities that the students could undertake when they came.

5.3.1.4   Time Factor as a Teaching Barrier

Debbie found that the limited time span available for her teaching session was a big challenge, especially when the schoolteachers requested her to address several topics in one lesson. For example, the GS School visit was focused on the Amazon Rainforest, which included three areas: plant adaptations, biodiversity conservation and rainforest music. She commented on this workshop:

> I wouldn't normally do a rainforest PowerPoint presentation and a music session at the same time. We were trying to do a huge amount in 45 minutes. …In my opinion it's too rushed and there's no time to explain anything properly. (Interview, July 2009)

The combination of the rainforest presentation and music was the schoolteacher's requirement and she was not very happy that the students did not have much time to practice their musical instruments, which were made of tropical plants:

> You need time to make good composition. Unfortunately they've got about two minutes to work on composition. They were all shy and embarrassed to perform in front of the rest groups as they are Year 8. (Interview, July 2009)

From this it can be seen that careful collaboration between the BGE and the schoolteachers would appear to be an influential factor if a productive school trip is to happen, given the limited time available for the visit.

5.3.2   The Structure of Debbie's Guided Visit

Both of the observed visits involved Debbie giving a 45 min PowerPoint presentation to the visiting schoolchildren inside the classroom and she did not accompany them on a tour around the garden on either occasion. However, during the initial fieldwork it was apparent that she did usually carry out guided tours after the formal talk, and during her interviews she explained that the objectives of her guided visits were (a) to develop the students' understanding of the rainforest environment, (b) to increase their awareness of deforestation in the Amazon region and (c) to appreciate indigenous culture by getting them to learn to play rainforest musical instruments. Talk played a dominant role during her presentations (see Fig. 5.5). The students from the GS School were able to compose music with the rainforest musical instruments and this was the only hands-on activity that they had. However, given the time limitations she was only able to show the rainforest musical instruments to the students from the WM School and they were not able to compose music.

The activities that the students from the GS School and the WM School had were mostly coded as passive directed, as they spent a large amount of the time listening to Debbie's presentation, although they were frequently challenged by questioning

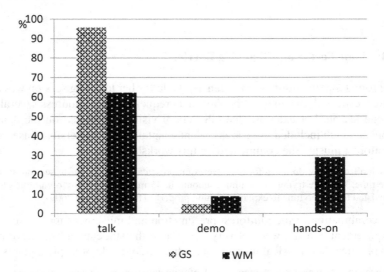

Fig. 5.5 The proportion of observed segments on Debbie's guided visits ($N_{GS}=45$ min, $N_{WM}=45$ min)

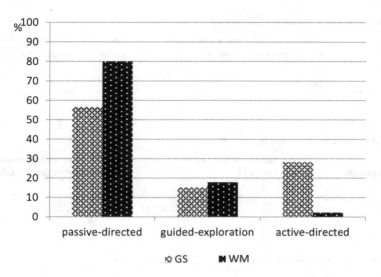

Fig. 5.6 Time spent on different types of learning activities during Debbie's guided visits (N_{GS}=45 min, N_{WM}=45 min)

Table 5.14 Types of learning activities during Debbie's guided visits

Types of activities	Activities
Passive directed	Listening to the introductory talk (discipline, health and safety, plan of the day)
	Listening to Debbie's presentation
	Watching the PowerPoint slides
Active directed	Practicing musical instruments
	Discussing with peers and composing rainforest tunes
	Performing to the class
Guided exploration	Reviewing previous knowledge on rainforests
	Guided discussion of plants and their use

(see Fig. 5.6). The guided-exploration and active-directed activities only played a peripheral role, as the students were only given a few minutes to perform such tasks (see Table 5.14).

5.3.3   The Discourse of Debbie's Guided Visits

From the analysis of the discourse data, it was found that 86 % of the lesson's discourse was contributed by Debbie and less than half involved interactions with students (see Table 5.15).

Table 5.15 Debbie's discourse by percentage

School groups	Percentage of Debbie's utterances in the lesson discourse (%)	Percentage of Debbie's discourse coded as interactive talk (%)
GS	86	43
WM	86	48

Table 5.16 Students' discourse by percentage

School groups	Percentage of student-initiated discourse (%)	Percentage of student discourse containing volunteering talk (%)
GS	4	3.5
WM	4	4

Table 5.17 Debbie's follow-up utterances ($N_{GS} = 126$, $N_{WM} = 168$)

School groups	Follow-up moves								
	Mnt (%)	Int (%)	Ect (%)	Evl (%)	Cfm (%)	Rpt (%)	Rev (%)	Prs (%)	Oth (%)
GS	1	42	13	24	2	12	6	0	0
WM	2	40	13	23	1	8	7	1	1

5.3.3.1 Student-Initiated Discourse on Debbie's Guided Visits

The contents of Table 5.16 show that student-initiated discourse was extremely limited, which was because the lesson was dominated by Debbie and probably also because when she did try to engage them they were too embarrassed to express their ideas in front of the whole class (Eder, Evans, & Parker, 1995; Eldredge, 2009).

5.3.3.2 Debbie's Responses to Students' Talk

Table 5.17 shows Debbie's preferred follow-up approach to students' talk, and it is notable that the 'insert' and 'evaluate' moves were predominant throughout the lesson discourse, which indicates that it was she who authorized and controlled the interaction with the students. As mentioned previously, she said during the interviews that she wanted all of the students to be active participants, but in practice only 13 % of the follow-up utterances could be coded as 'elicit', these being those that have the potential to promote student contribution to the construction of shared knowledge.

The discourse below shows an example of her follow-up approach, in which she challenges the students to surmise whether epiphytes need moisture and nutrients to grow.

Transcript			Move
1	Debbie:	The roots normally absorb water and nutrients	
2		But epiphytes don't have roots	
3		So do the epiphytes need water and nutrients?	PE
4	Ss:	No	
5	Debbie:	No?	Mnt
6		Are you sure?	Mnt
7		They are living things, aren't they?	Tag
8		Don't all living things feed with water?	PE
9	S1:	They don't have roots	
10		But they can get water from the plant they grow on	
11	Debbie:	Great thinking	Evl
12		Unfortunately it's not how it works	Evl
13		But some plants like mistletoes get water from the tree they grow on	Int
14		Look at this old man's beard plant, it's an epiphyte	Int
15		When it rains the leaves suck water	Int
16		Meanwhile dead animals and dead leaves might drop in as well	Int

Here, many of the students think that epiphytes do not need water and nutrients for living and so respond 'No' together (utterance 4). Debbie repeats the answer with a questioning voice to challenge their thought (utterance 5), expecting someone to justify their reason for saying 'No'. She goes on to prompt the students by offering some clues (utterances 7–8), and then S1 suggests that even though the epiphytes do not have roots, they could absorb water from the host plant (utterances 9–10). S1's answer is not correct as she confuses epiphyte with parasitic plants, such as mistletoe, which grows on the host plant (utterances 11–13). Realizing that the children are unable to give a better answer, she explains how an epiphyte gets water and nutrients from the environment (utterances 13–16). Although Debbie's cuing did not help the students to provide a correct explanation about the epiphyte's adaptation, S1's reasoning helped her to realize that some students were confused between epiphytes and parasitic plants. This would suggest that the 'elicit' and 'maintain' moves not only promote children's thinking but also assist educators in that it gives them practice in ensuring that their questioning is appropriate and pitched at the right level for learning to occur.

5.3.3.3 Debbie's Questions

Table 5.18 shows that the majority of questions that Debbie asked were 'right answer' questions, which only allow for certain predetermined correct answers. The pervasiveness of such questions in her teaching practice could have reflected the

Table 5.18 Types of questions proposed by Debbie ($N_{GS} = 48$, $N_{WM} = 56$)

School groups	Open ended (%)	Right answer (%)	Procedural and tag (%)	Invite participation (%)	Other (%)
GS	18	49	27	4	2
WM	20	50	26	4	2

schoolteachers' expectations of the visit, for as Debbie explained during one interview, the schoolteachers from the GS School and the WM School had expressed the hope that the visits would reinforce the students' knowledge of the Amazon Rainforest that they had acquired in the school classroom.

The discourse below shows an example of a 'right answer' question posed to students, where Debbie starts the introduction to rainforest conservation with a mathematics question, asking them to calculate the amount of trees cut down in the rainforests in 45 min.

Transcript			Move
1	Debbie:	We know about 2,000 trees are cut down in rainforests around the world in a minute	
2		Here is a mathematics question	
3		By the time when we finish we have been here for 45 min how many trees will be cut down in the rainforest?	RA
4	S10:	90,000	
5	Debbie:	Brilliant	Evl

Rather than challenging the students' mathematical skills, the purpose of the question in the above discourse was to make the students aware of the startling fact about the number of trees that were going to be cut down during the course of the lesson, so she could move on to discuss the importance of the need to protect the Amazon Rainforest.

5.3.3.4   Students' Questions on Debbie's Guided Visits

There was only one student-generated question put during the discourse with the GS School group and none came from the UW School group. The single student-generated question was a question of a procedural nature, where one student asked how long they could play the rainforest musical instruments for.

5.3.4   Students' Views on Debbie's Guided Visit

The students had various perceptions of the visiting experience and Debbie's teaching style according to the impression sheets they completed subsequent to the visit (return rate was 50 %, N = 80). Moreover, it turned out that for a significant number the botanic garden visit was a brand new experience, with 30 reporting that it was their first time.

There were 11 students who expressed that they enjoyed the visit inside the tropical glasshouse most, with the humidity and hot environment and the exotic plants displayed inside the glasshouse being given as their most memorable impressions. For example, one student wrote:

> I thought the most impressive part of the trip was the rainforest glasshouse because we saw all the different types of plants and we got to experience the heat.

Interestingly, some students connected the tour around the tropical glasshouse with their previous experience:

> When we went into the tropical room, it was humid and hot. It reminded me of my holiday to Jamaica.

Moreover, given that the rainforest is a remote environment for the students living in England, the trip to the botanic garden gave them an opportunity to experience the rainforest without travelling afar:

> Visiting the rooms where we could experience the temperature and humidity of the rainforest was the most interesting part of the visit. Because if you never have been to a place with a tropical climate you know roughly what it's like.

However, as Wandersee (1986) reported, young children usually find animals more interesting than plants and nine students stated that the most interesting part of the trip was seeing the different animals, such as the peacocks and parrots. For example, in this regard, in responding to the question 'what is the most interesting part of the trip?' a student wrote:

> The wildlife [that I saw on the visit was the most interesting part]. Because the parrots could talk and the peacock did different things like show off its feathers.

Moreover, a couple of students reported that they liked the section when their schoolteacher allowed them to explore the garden by themselves. This supports Falk and Dierking's (2000) argument that learning is driven by the unique intrinsic needs and interests of the learner when free choice is given.

Compared to the above comments on schoolteacher-led sessions, only nine students viewed Debbie's presentation as the most interesting. Among those respondents, five of them reported that her presentation was informative and that some of her explanations were full of humour. Four students wrote that they had enjoyed playing the rainforest musical instruments, although they regretted that they did not have much time to compose their rainforest tunes.

Although the majority of the students stated that they enjoyed the visit, several of them expressed the view that some parts of the trip were quite disappointing. For instance, 13 students complained that the BGE-led session was too long. A few students also complained about the facilities in the botanic garden, citing things such as the slippery floor in the glasshouse, too few birds in the bird house and the heat inside the classroom.

In addition, Debbie's workshop was highly rated by the students according to what they wrote on their impression sheets, with their reporting that they learned a lot of new and interesting facts from her presentation, for example:

> The toothpaste and hair gel are made out of some plants.
> Bamboo is the quickest growing plant in the world. It grows a meter per day.
> Only 1 % of rainforest plants have been tested for medical use.

As discussed earlier, the teaching strategies that she employed, such as asking questions and inviting the students to the front of the class to do demonstrations, had great potential to facilitate learning engagement and participation. The students gave her positive feedback on this interactive teaching approach:

> [I liked] The way she got my class involved with what she was doing because seeing a friend doing something that you might not be interested in makes you interested.

However, some students complained that the presentation was too long and some parts overlapped with what they had learned at school. Nevertheless, many of the students wrote that they preferred to see the botanic garden, rather than sitting inside the classroom learning about the same subject matter. One student even suggested that the botanic garden should provide them with booklets of information so that they could spend more time exploring the site, rather than sitting inside the classroom listening to the BGE's talk.

5.3.5   Case Summary

As Debbie explained during interview, the purpose of the classroom-based presentation was to inspire curiosity and stimulate interest, whereby the students would become self-motivated to explore the botanic garden. Moreover, the guided visits were informative and cross-curricular, integrating art, geography and science. The rainforest musical instruments session during the GS School visit gave the students an opportunity to appreciate the indigenous culture. Further, the performance on these musical instruments contributed to the development of the students' creativity and teamwork spirit. Although the presentation of rainforest conservation was not requested by the schoolteachers, Debbie expressed the belief that it was necessary for the students to understand the growing importance of this issue, and in support of this view, on the impression sheets many students stated that the most fascinating fact they found on the visit was the speed of the loss of rainforest. In addition, a lot of the children reported that they would like to learn more about plant conservation, for instance, writing:

> How can I prevent trees and plants being cut down?
> Why are people cutting down the rainforest?
> How long it will be till all the rainforests disappear if deforestation carries on like it is now?

The analyses of the discourse data have shown that the students were not active participants during the whole class discussion, although Debbie did attempt to get them involved through questioning. However, even if she had tried to engage the groups more through, for example, asking open-ended questions to prompt their thinking, because these were adolescents who would be scared of losing face if they got it wrong in front of their peers, it is unlikely she would have fared any better had she tried even harder. Research has shown that adolescents' perceptions of teacher support, such as the teacher promoting interaction and mutual respect, have a positive influence on their motivation and engagement (Ryan & Patrick, 2001). In this regard, the young adults should be given more encouragement and support from the educators so as to promote learning participation.

5.4 Case: Julia

Julia was the only full-time education officer for visiting school groups in the Garden C, and at the time of data collection, she had been working as a BGE for three and a half years. She was the only research participant who has a doctorate degree in botanical science. Besides her PhD research on woodland plant ecology, she had also worked as a horticulturalist in the 'Rose Garden' of Regent's Park in London for a year. As she found encouraging children to learn about growing plants interesting, she later became a primary teacher, working first with lower primary-aged children and, subsequently, with children with special needs. The teaching experience in mainstream schools and special schools allowed her to be a confident outdoor educator, in terms of designing, organizing and delivering education programmes to school groups.

5.4.1 Julia's Perceptions of Teaching and Learning Science Outdoors

According to the interview data, Julia's pedagogical beliefs could be summarized as 'preparation as the key to a successful visit', 'exploring as a scientist', 'maintaining engagement through small group teaching' and 'learning about plants and environmental issues'. Moreover, she believed that a teacher's job should be used to enthuse and encourage children to be amused and interested in the world around them.

5.4.1.1 Preparation as the Key to a Successful Visit

Research on school trips to informal settings has found that effective pre-visit preparation has a positive influence on visitors' learning experience (Falk & Dierking, 1992; Hargreaves, 2005; Orion & Hofstein, 1994). Julia believed that for the school trip to a botanic garden, the schoolteachers' involvement in preparation and planning work should not be limited to the discussion of the visiting schedule with the

BGE, but should also be concerned with the students being informed about the purpose and the content of the trip. In order to support schoolteachers in preparing students for visits, she had designed a series of PowerPoint slides, which introduced the history of the garden, the activities on the trip and the facilities available. She also suggested the use of Google Earth™ as a tool to introduce the trip to students. However, only a small number of schoolteachers gave their students an orientation and preparation talk prior to the visit. Julia argued that if they were prepared in advance, they could better concentrate on tasks relevant to learning:

> I really would like to feedback to the schools to say if you'd done x, y and z before you came, you would have a better visit, at least the children would ask more high quality questions. (Interview, April 2009)

In this regard, in their study on school trips to museums, Griffin and Symington (1997) found that most teachers did not prepare students for visits and used task-oriented teaching practices, which failed to connect school topics to the museum context. Orion and Hofstein (1994) also suggested that proper pre-visit preparation can minimize the novelty of a school trip experience and yet still stimulate children's interest.

5.4.1.2 Exploring as a Scientist

Julia expressed the opinion that children learn best when they are free to explore, because when they are able to have sensory experiences, such as touching, smelling, hearing, seeing and tasting, they would be engaged in thinking and meaning making. She emphasized that the physical experience during first-hand data collections plays a significant role in scientific research and thus, she highly recommended that the students were given opportunities to explore by using their sensory modalities:

> My big ethos about getting children to come here is to encourage them as scientists to look and scientists need to record things by drawing. Then drawing is a very important skill that scientists will use to make observations. (Interview, April 2009)

She argued that drawing is not only a format that engages students through observation but also a process of promoting productive thinking. For example, she expected them to ask questions when they were drawing desert plants, such as 'why does the leaf shape look like that?' In this regard, Dempsey and Betz (2001) suggest that drawing is an important skill for children to demonstrate their biological observations, and rather than only describing the objects, Brooks (2005) argued that it provides children with 'a visible mediating role for communication, meaning making and problem solving' (p. 82). That is, drawing not only records students' observations but also develops their thinking.

In order to encourage students to be actively engaged in enquiring, Julia preferred not to use worksheets on her guided visits, because she thought that they were a distracting instrument and may limit engagement. In responding to the question of how to engage students with the visit, she stated that:

> They physically do things in school, a lot with pen and paper. I try not to have any pen and paper when they are here unless it's drawing. (Interview, June 2009)

In fact, a lot of activities during her guided visits required the students to be hands-free, including pond dipping, minibeasts hunting and plant collage. These hands-on activities offered the students the chance to interact with the natural world, which stimulates their interest in exploring and investigating. Julia explained that for most of the children, it was their first time visiting a botanic garden, and thus, it was better to get them enthused, rather than spoiling their enjoyment by focusing on one specific investigation. However, for the school groups who visited the botanic garden frequently, she suggested that the activities should be more enquiry based and focus on a specific topic that fitted in with an ongoing school study.

5.4.1.3 Maintaining Engagement Through Small Group Teaching

Teaching outside the classroom is different from inside, where students can easily lose their concentration; thus, the educator needs to pay careful attention to organizing the logistics of the visit. In practice, Julia usually divided the visiting school group into several subgroups to guarantee a small group for instruction (usually less than 15 students in one group). In her words, the experience for the children would be improved if they had more assistance from adults in the group. In order to provide this ratio of children to expert adult, she had set up a group of volunteers who had been trained to be the 'expert helpers' with the small groups.

When responding to the issue of classroom interactions, she argued that the small group instruction enabled each student to have the opportunity to talk to the educator or accompanying adults:

> I think the interaction is very important and that's one reason I try to have them in small groups with an adult who can ask questions and the children are encouraged to ask questions. (Interview, June 2009)

From her point of view, the dialogue between adults and children on the trip was critical for learning engagement and, hence, a meaningful experience. That is, she believed that the adults' role during the visit assisted the level of children's learning:

> So having somebody who has some authority and experience is important for children's learning. (Interview, June 2009)

5.4.1.4 Learning About Plants and Environmental Issues

For most of the visiting school groups, the schoolteachers requested that her teaching be focused on the science that fitted in with what they were studying at school, but she always endeavoured to give the students a broader view about plants and how important they are to human lives:

> I try to make the program have other aspects. It includes the most popular topics, such as: climate change, sustainability, biodiversity and fair-trade. (Interview, June 2009)

She suggested that the design of education programmes for school groups should meet the requirements stated on the National Curriculum and the working agenda of the BGCI, which emphasizes promoting biodiversity and conservation. This view is consistent with that of researchers, who have contended that the botanic garden visiting experience could contribute to the improvement of one's environmental awareness (Ballantyne, Packer, & Hughes, 2008; Morgan, Hamilton, Bentley, & Myrie, 2009).

5.4.2   The Structure of Julia's Guided Visits

According to the observational data (see Fig. 5.7), Julia devoted most of her effort to presenting the information about the flower parts and their functions through 'talk'. Unfortunately, the arrival of students from the UW School was delayed 1 h due to the traffic, and in the later interview she acknowledged that the late arrival had affected her plans, because she did not have much time left to get them to do hands-on activities.

The coding of the types of learning activities (see Table 5.19) shows that she performed as a facilitator as a large proportion of the observed visits were based on guided exploration (see Fig. 5.8). Students were passively motivated when they were listening to her instruction, such as the demonstrations on how to do pond dipping and construct a flower model. However, they were actively engaged for most of the time when they were dipping in the pond, walking in woodland and observing plants in the glasshouse.

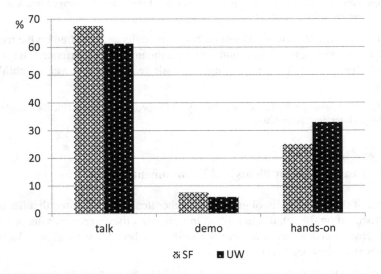

Fig. 5.7 The proportions of observed segments during Julia's guided visits ($N_{SF}=85$ min, $N_{UW}=40$ min)

Table 5.19 Types of learning activities during Julia's guided visits

Types of activities	Activities
Passive directed	Listening to the class management instructions (especially about the discipline)
	Listening to the BGE's demonstration (pond dipping, flower model)
	Listening to the BGE's lecture (features of different living environments)
Active directed	Drawing whilst observing
	Asking questions of the BGE or schoolteacher
	Hands-on (pond dipping, collage)
Guided exploration	Guided discussion of flower structure
	Touching/smelling plants when permitted
	Exploring the glasshouse/garden with the BGE's guidance

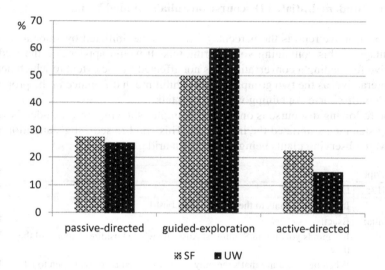

Fig. 5.8 Time spent on different types of learning activities on Julia's guided visits (N_{SF}=85 min, N_{UW}=40 min)

5.4.3   The Discourse of Julia's Guided Visits

According to the analyses of the discourse between Julia and the students (see Table 5.20), her talk predominated during the lesson discourse. Moreover, it emerged that there was a big difference in her discourse coded as interactive talk between the two observed groups, so it would seem that the limited teaching time for the UW School group forced her to be less interactive when communicating with the students.

Table 5.20 Julia's discourse by percentage

School groups	Percentage of Julia's utterances in the lesson discourse (%)	Percentage of Julia's discourse coded as interactive talk (%)
SF	82	54
UW	86	32

Table 5.21 Students' discourse by percentage

School groups	Percentage of student-initiated discourse (%)	Percentage of student discourse containing volunteering talk (%)
SF	23	40
UW	18	36

5.4.3.1   Student-Initiated Discourse on Julia's Guided Visits

Table 5.21 above reports the percentage of discourse initiated by students and the percentage of this containing volunteering talk. It would appear that the students' initiative for leading a conversation was not affected by the extent to which her talk was interactive, as the two groups did not exhibit much difference in the proportion of student discourse containing volunteering talk.

The following discourse is one of the examples showing how a student's volunteering statements initiated the interaction. This child raised Julia's attention when they were observing plants being grown in an arid glasshouse.

Transcript			Move
1	S13:	Look!	VoL
2		It's burned [points to the leaves on the bush]	
3	Julia:	Exactly	Evl
4		It reminds you that in Southwest Australia where bushfires happen all the time	Int
5		When there is a fire that's the only time those seed cases will open to let the seeds out	Int
6		So to get seeds to grow new plants there has to be fires every few years	Int
7		So to get seeds to grow new plants there has to be fires every few years, although from human's point of view is bad news	Int
8		From the plants' point of view they've developed over thousands, millions of years to get ready to do that	Int

The burned leaves on the bush that S13 has found during his observations have stimulated his curiosity, and his comment raises Julia's attention, who then provides much information to explain the student's finding (utterances 4–8). However, although the student's realization that the bush leaves had been burnt was thought provoking and he could well been wanting to know why this particular bush was burnt, Julia's several responses did not pick up on that aspect. That is, it would appear that her utterances had the effect of shutting down any potential dialogue.

Table 5.22 Percentage of Julia's follow-up utterances ($N_{SF}=258$, $N_{UW}=71$)

School groups	Follow-up moves								
	Mnt (%)	Int (%)	Ect (%)	Evl (%)	Cfm (%)	Rpt (%)	Rev (%)	Prs (%)	Oth (%)
SF	3	68	3	14	3	5	2	0	2
UW	1	73	0	17	1	4	3	0	1

5.4.3.2 Julia's Responses to Students' Talk

The coding of follow-up utterances shows that Julia controlled the exchanges of ideas as the 'insert' move predominated in the follow-up discourse (see Table 5.22). As discussed earlier, direct responses to student's talk will not generate more contributions from students, but rather tend to terminate the interaction.

The discourse below shows the pervasiveness of Julia's direct feedback to the students' talk. When Julia is explaining about carnivorous plants needing to digest insects for nutrients to survive in boggy areas, a student asks:

Transcript			Move
1	S11:	To keep these insects alive, do you have to feed them with insects?	VoL
2	Julia:	What we actually do is to put larvae around	Int
3		Sometimes there are flies around as well	Int
4		It's important you can't force them to eat otherwise you just kill them	Int
5		For the Venus flytraps the leaves only close three or four times	Int
6		Then the leaves will die	Int

The question asked by S11 in the above discourse shows his curiosity about the horticultural perspective of carnivorous plants (utterance 1). Julia offers detailed information to respond to the student's question (utterances 2–6). The insert move did prohibit the development of the conversation; however, the immediate responses probably gave the students a feeling that the educator respected their contribution and hence, they felt feel free to initiate more questions or comments. In some circumstances, however, an educator has the responsibility of promoting student thinking, and Julia did this on occasion by prompting the students for ideas.

Transcript			Move
1	Julia:	Who can tell me what compost is?	PD
2	S7:	It contains plants and flowers in it	
3	Julia:	It can be like a growing medium, something to make things grow from	Mnt/Ect
4		Can anyone else tell me what compost is?	PD
5	S2:	It's made of leaf litter	
6	Julia:	It's not just leaf litter	Evl
7	S9	It has things like dead leaves	
8	Julia:	And dead plants, dead bits of old apple cores	Cfm/Int

She initiates the conversation with the students with a question 'Who can tell me what compost is?' (utterance 1) and S7's response 'It contains plants and flowers in it' (utterance 2) is not clear according to her perspective. So she rephrases S7's response so that the rest of the class can hear (utterance 3). Then in order to have other students' ideas heard, she holds back the answer and asks 'Can anyone else tell me what compost is?' (utterance 4). As S2 and S9's responses are not complete, Julia adds further information to develop a full definition of compost (utterance 8). This discourse shows that the 'elicit' and 'maintain' moves can contribute to the co-construction of knowledge between students and the educator working together.

5.4.3.3   Julia's Questions

Table 5.23 shows that the proportions of open-ended questions that Julia asked during the SF School and the UW School visits varied, thus indicating that she adapted to the specific conditions of each school group. However, the greater proportion of 'right answer' questions posed to the students from the UW School may have related to the shortened visiting time, for when compared to closed-ended questions, open-ended questions are less structured and require longer time to prompt students' thoughts.

5.4.3.4   Students' Questions on Julia's Guided Visits

Table 5.24 presents the percentage of student-generated questions, coded as 'higher order' and 'lower order', by the students from the two visiting school groups, and it can be seen that those in the SF School group proposed more higher-order questions than the other group. In this regard, during the SF School's guided visit, it was observed that Julia adopted a more interactive approach to communicate with the students and asked a higher proportion of open-ended questions and consequently,

Table 5.23 Types of questions proposed by Julia ($N_{SF}=65$, $N_{UW}=13$)

School groups	Open ended (%)	Right answer (%)	Procedural and tag (%)	Invite participation (%)	Other
SF	45	11	37	5	2
UW	8	47	40	3	2

Table 5.24 Types of questions generated by students ($N_{SF}=23$, $N_{UW}=6$)

School groups	Lower order (%)	Comprehension (%)	Predict (%)	Synthesis (%)	Evaluate (%)
SF	65	26	9	0	0
UW	83	17	0	0	0

her teaching with this group was more supportive than with the other, which could explain this difference between the groups. That is, perhaps the students were able to put better quality questions, because she created a more proactive role for that particular class.

5.4.4 Students' Views of Julia's Guided Visits

There were 59 copies of impression sheets sent to the visiting schools observed and 54 copies were sent back 1 week after the visit. According to the responses on them, most of the students had visited the botanic garden many times and only ten of them had been there less than three. It is relatively unusual for children who come on school visits to have been before and very few have been several times. The SF School, however, is an independent school which is within walking distance of the Garden C and uses the site extensively, each year having many of its students come on school visits. In addition, owing to the school being local, the parents would often bring their children to the Garden C for leisure purposes at weekends, as well as at the end of the school day.

A large number of students found the visit was interesting and full of fun, and there were 21 students who responded that they thought the hands-on activities, such as pond dipping, minibeast hunting, observational drawing and collecting plants from the grounds, were the most interesting parts of the trip. For example:

> I think it was the pond dipping that was most interesting because you had to dip your net into the water and see what you could find.
> The most interesting session was the compost [mini-beast hunting] because I got to touch things.
> I liked drawing cacti most because it involves doing something about plants on paper.

Research has suggested that an activity that promotes multisensory engagement has a greater potential to facilitate the learner's meaning making (Zubrowski, 2009). Consistent with this claim, a number of students reported the concepts they had acquired from the participation in the hands-on activities in the form of the following drawings (see Fig. 5.9), each of which illustrates the acquisition of new knowledge.

In Frame A, the student is describing pineapple plants as 'pineapples sat on top of plants and not on trees'. Perhaps it was the first time he had seen a growing pineapple and it changed his previous conception that pineapples grow on trees. Frame B depicts how a dragonfly larva swims in water by absorbing it and squirting it out from its rectum. In sum, it would appear that these students were highly engaged with their observations and that the revelations contributed to the development of their interest in learning ecological science. Moreover, they themselves reported that these effective learning opportunities were their favourite, which probably explains their high levels of motivation and this reinforces the above researchers' argument that multiple sensory activities are very important learning tools.

Frame A   Frame B

Fig. 5.9 Drawings reporting the most interesting things students found on the trip

Apart from astounding facts, a few students expressed that the aesthetic and exotic perspectives of the plant kingdom interested them most. One girl from the UW School wrote that:

> I liked the plants in the tropical glasshouse most because they are so beautiful.

Another student reported that he enjoyed the tour round the tropical glasshouse most as he had the freedom to explore the plants:

> The most interesting part was the tropical glasshouse because you were free to look and walk around.

Even though the majority of the students stated that they enjoyed the visit, several pointed out that some parts it were not very interesting or even boring. This was probably because young children nearly always like to do something new and as many of them had been before, they were not very motivated to join in. For instance, some found the minibeast hunting was not stimulating:

> The least interesting part was searching through the compost because we have done it all before.

A couple of students complained that some tasks were not engaging and thought they did not learn anything. For example:

> Walking around the place looking at plants [was boring] because it was quite plain.
> Drawing the cacti in the arid house [was boring because] we didn't learn anything and we didn't have any time to look at the cacti.

One of the sad facts of life about being in a school is how much time has to be spent waiting for everyone to do something, such as queuing for the toilet. In this regard, some students complained that they had spent too much time on off-task activities, rather than exploring the botanic garden.

A high proportion of the students rated Julia as a knowledgeable educator who knew a lot about plants and animals, with 49 students expressing the view that she was a good informant about botanical science as she gave plenty of facts. In responding to the question 'What do you like about the lesson led by the garden teacher?' some students wrote:

> She taught me lots of interesting facts I didn't know.
> She gave you lots of facts about the plants and she made you want to learn more.

Fig. 5.10 Student's
perception of Julia's
teaching

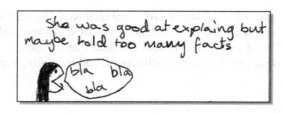

The students' views on Julia's teaching might be related to her approach to interacting with them. In this regard, a large part of Julia's initiated discourse and 'insert' moves determined that the content of the talk was full of declarative facts and there was only one student who thought that she presented too many facts to them (see Fig. 5.10).

At interview, Julia explained that in the botanic garden environment where so many things are new and different, the schoolteacher must 'tell' facts that make connections which children would not have time to see. Moreover, she expressed the belief that there is something intrinsically different about 'teaching as telling' in an outdoor environment to school, because the students are also being asked to look, feel and generally understand what they are experiencing. Having only limited time to observe plants in the glasshouse did not satisfy a few students, with one writing:

> It was really interesting to look at the plants in the glasshouse but she pulled you away from things too quickly which didn't give us much time to look properly.

Given the time constraints mentioned above, Julia could only give them a very short time to explore the glasshouse and a further reason for this was because their schoolteacher had asked her to cover a range of different topics, so she had to keep to the originally agreed schedule.

5.4.5 Case Summary

In the interviews Julia stated that her guided visits were designed to develop students' interest in learning ecological science and helping them explore as scientists, some of which she made possible by engaging them in doing hands-on activities. However, most of the activities were carried out through guided exploration rather than allowing the students to investigate by themselves. Garden C does not allow children to be unaccompanied by adults, in part as a safety measure, but mainly so as to protect the fabric of the garden from the children and the kind of depredation they could cause, if they were able to move about unsupervised, 'en masse'.

Julia also emphasized the significance of having small groups so as to facilitate interactions with the students; however, her immediate response to students' talk by providing them with abundant information normally failed to extend the conversation with the students. Moreover, the analysis of the discourse data has found that the length of the guided visit may have had an impact on the quality of questions that Julia was able to ask. That is, it would appear that the longer the teaching time available, the more extended the BGE-student dialogue through questioning that could take place. It also emerged that the quality of the students' questions was influenced by how much support and prompting they received from Julia herself.

5.5   Cross-Case Conclusion

In the case studies reported above, the pedagogical practices of the participating BGEs were analysed from four perspectives: perceptions of teaching and learning in informal contexts, structure of guided visits, approaches regarding communicating with students and students' views of their visit experience. In this section, these findings are addressed through taking a cross-case perspective, in which each of the participating BGE's pedagogical practices is compared.

5.5.1   The Professional Roles of a BGE

All four participants emerged as having distinct identities as BGEs, in spite of their having several roles in common, such as being a garden historian, a manager of children and adult visitors and a facilitator of learning. More specifically, these BGEs can be categorized as performing the following roles: botanical guide, science educator, environmental educator and botanist, each of which is discussed below.

The key role that Mark performed during his teaching was that of botanical guide. First of all, he agreed that a guide needs to be familiar with the venue where she/he works in. His 15 years of working experience in Garden A had enabled him to understand how to use the garden facilities and plant collections to organize school visits in an effective and efficient way. For most children, plants are boring when compared to animals and other objects (Wandersee, 1986) and in order to increase student interest, he employed different interpretive approaches, such as the use of metaphors, analogies and storytelling so as to explain aspects of ecological science. According to their impression sheet responses, many students commented that his explanations were interesting, simple and memorable. The ways in which he interacted with students revealed his role as a guide, more specifically, rather than posing questions, he preferred to listen to those generated by the students and then respond by giving detailed explanations.

A focus on learning the language of science was the most prominent feature of Simon's guided visits. Although the acquisition of language and literacy has been promoted for decades in science education, as an experienced science teacher, he had come to believe that the rote teaching carried out in schools fails to help students use the language of science, because they tend to just memorize sets of scientific vocabularies. This point was endorsed by Lemke (1998) when he argued that 'the one single change in science education that could do more than any other to improve student's ability to use the language of science is to give them more actual practice using it'. The analysis of Simon's case indicates that learning outside of the classroom is a good way of filling in the gaps of school education as well as reinforcing previous learning. To this end, it was observed that he exploited, effectively, questioning strategies, such as the verbal jigsaw and the reflective toss, during his teaching.

Debbie was the only participating BGE who discussed plant conservation issues during the guided visit, even though according to the mission statement of the BGCI, environmental education is an important dimension of botanic garden education. Although the content of the guided visit was largely determined by school-teachers' requests, Debbie integrated the conservation issues into her teaching and thus kept to the working agenda of the botanic garden, namely, to promote biodiversity conservation matters to its visitors.

On her guided visits, Julia introduced a number of scientific names of plants, for example, *Sarracenia* (pitcher plant), *Drosera* (sundew) and *Dionaea muscipula* (Venus flytrap). Arguably, students should be introduced to the scientific names of the plants (Sanders, 2004). However, Julia only told the students the scientific terms for plants and did not explain the rationale for naming plants in Latin, that is, unlike Mark, she failed to enlighten them on the meanings contained within in plant taxonomy.

Taking the four above roles together, namely, botanical guide, science educator, environmental educator and botanist, it is posited that an ideal BGE should be able to perform all of these in their practice. With this perspective, along with that developed regarding dialogue and questioning in the literature review, in Table 5.25 below the strengths of each BGE are listed as well as areas where they could improve their delivery.

Table 5.25   Reflections on the observed teaching practice of the BGEs

	Strengths	Perspectives that could be improved
Mark	Familiar with the garden facilities and plant collections	Few BGE-initiated questions
	Used different explanation strategies: metaphors, analogies and storytelling	Environmental education was not directly integrated into the visit
	Responded to students' questions in detail	The activities did not focus on talking science
	Promoted learning through sensory modalities	
Simon	Employed different questioning strategies	Less knowledge about botany
	Encouraged students' thinking through exploration	Few student-generated questions
		No connection with environmental education
Debbie	Incorporated environmental education within the guided visits	Presentation of a passive-directed learning activity
	Integrated different school subjects into one workshop	No student-generated questions or comments
Julia	Expertise in botany	Teaching through telling
	Exploration through hands-on activities	

5.5.2   The Structure of the Guided Visit

The analysis of the observed guided visits' structure in the above case studies has presented the individual characteristics of each participating BGE's approach to organizing lessons for the different visiting school groups. It is noted that the activity coded as 'talk' took a larger proportion of the time in Debbie and Julia's lessons as compared to Mark and Simon's, which suggests that the latter two tended to give students more opportunities to use their sensory modalities to experience the natural world, whereas the two former addressed ecological science issues mainly through their verbal communication.

With regard to the extent of student activities, it is apparent that these were encouraged by most of the BGEs; with the exception of Debbie, the observed guided visits mostly involved PowerPoint presentations. As reported earlier, some students complained in their feedback that sitting and listening to Debbie's lecture was not as interesting as exploring the plants in the glasshouses by themselves. 'Talk' was also greatly observed during Julia's guided visits; however, much of it was devoted to the guidance of students' work or activities, for instance, making observational drawings, pond dipping and so forth. It emerged that the structure of the learning activities during the visits affected the students' level of engagement, in that even though during her presentation she tried to involve the students by using questioning, the lecture form was not conducive with encouraging a group of self-conscious adolescents to participate, because they were unlikely to subject themselves to potential ridicule from their peers.

5.5.3   The Discourse on the Guided Visits

The case analysis has shown that students' talk only took up a very small proportion of the lesson discourse. However, they did employ volunteer talk at certain times, and it is notable that the students in Mark and Julia's groups were more active than those in Simon and Debbie's in initiating conversations. From the analysis of the student-generated questions, it also emerged that Mark and Julia's students asked more higher-order questions.

It is noted that the students who were guided by Mark and Julia were all from independent schools, whilst the students in Simon and Debbie's groups were from state schools. According to the Independent Schools Council's (2010) statistics, the teacher to student ratio in the private sector is 1:11, compared to 1:17 in the state sector. Therefore, the high teacher to student ratio in independent schools allows their students to get the attention they need much easier than the ones attending state schools. Moreover, under such circumstances teacher-student dialogue is more likely to occur than with larger groups, because the essential quality time needed for this is available.

Table 5.26 BGE responses to student contributions

	Mark		Simon		Debbie		Julia	
	FB (%)	BS (%)	SB (%)	NP (%)	GS (%)	WM (%)	SF (%)	UW (%)
Promoting dialogic interactions	17	13	47	72	32	36	15	9
Demoting dialogic interactions	83	87	53	58	68	64	85	91

The above individual case analyses have shown that the nature of BGE-student interaction was largely influenced by how the BGEs responded to the students' discourse. In Table 5.26, information is presented regarding the extent to which the discourse on the guided visits was dialogic in nature. The proportion of follow-up moves that facilitated dialogic interaction ('maintain', 'elicit', 'revoice', 'repeat' and 'press') by each BGE was not as great as that for the moves that demote dialogic interaction ('insert', 'evaluate' and 'confirm'). According to the data, Simon and Debbie more often than Mark and Julia promoted the students' thinking by having them contribute to the discourse, whereas the latter two had a greater tendency to give the students direct explanations. This finding may relate to the level of each BGEs' pedagogical knowledge, in that, as explained earlier, only Simon and Debbie had teaching qualifications and the formal teacher training experiences may have equipped them with a stronger awareness of the importance of involving students in class discussions than for the other BGEs. The students' willingness to engage in volunteering talk may also have had an influence on the BGEs' choices regarding offering feedback, in that as there were less such contributions from Simon and Debbie's students and thus, these two BGEs had to devote more effort to eliciting ideas from their students. With respect to this, it is posited that although direct instruction is useful for delivering factual information, it might not be an effective strategy for contributing to students' conceptual understanding.

Making a distinction between the open-ended and the closed-ended questions reveals the extent to which the BGEs' questions were supporting learning by challenging the students' thinking. All the BGEs posed more closed questions than open-ended questions to their students. Only in Mark's case did the open-ended questions exceed 45 % of the total, whilst for the other BGEs there were two or three times more closed-ended questions than open-ended ones. From this it can be deduced that the participating BGEs, with Mark being the exception, were not effective in scaffolding or pushing forwards the students' thinking during their lessons. However, as DeWitt and Hohenstein (2010) have argued, the demands placed on the educators by the different contexts, for example, the students' backgrounds, the size of class and the requirements set by the schoolteachers, affect the questioning strategies that they can employ in their practice.

In this chapter, the pedagogical practices of each participating BGE have been presented through consideration of the multiple cases. That is, for each case study the perception of the BGE regarding the teaching and learning in the botanic garden context, the structure of the guided visit, the interaction between the BGE and students and the students' views of their visit has been discussed. In the next chapter the pedagogical behaviours that the participating BGEs demonstrated during their practice are analysed in order to address the third research question of this study, namely, what were the pedagogical behaviours that the participating BGEs exhibited when engaging with and supporting the children's learning?

References

Ballantyne, R., Packer, J., & Hughes, K. (2008). Environmental awareness, interest and motives of botanic gardens visitors: Implications for interpretive practice. *Tourism Management, 29*(3), 439–444.

Barratt, R., & Barratt Hacking, E. (2008). A clash of worlds: Children talking about their community experience in relation to school curriculum. In A. Reid, B. B. Jensen, J. Nikel, & V. Simovska (Eds.), *Participation and learning: Perspectives on education and the environment, health and sustainability* (pp. 285–298). London: Springer.

Bloom, B. S., Engelhart, M. B., Furst, E. J., Hill, W. H., & Krathwohl, D. R. (1956). *Taxonomy of educational objectives: The classification of educational goals* (Vol. 1). New York: Longman.

Brooks, M. (2005). Drawing as a unique mental development tool for young children: Interpersonal and intrapersonal dialogues. *Contemporary Issues in Early Childhood, 6*(1), 80–91.

Chin, C. (2007). Teacher questioning in science classrooms: Approaches that stimulate productive thinking. *Journal of Research in Science Teaching, 44*(6), 815–843.

Csikszentmihalyi, M. (1975). *Beyond boredom and anxiety*. San Francisco, CA: Jossey-Bass.

Dempsey, B. C., & Betz, B. J. (2001). Biological drawing: A scientific tool for learning. *The American Biology Teacher, 63*(4), 271–279.

DeWitt, J., & Hohenstein, J. (2010). School trips and classroom lessons: An investigation into teacher-student talk in two settings. *Journal of Research in Science Teaching, 47*(4), 454–473.

DeWitt, J., & Osborne, J. (2007). Supporting teachers on science-focused school trips: Towards an integrated framework of theory and practices. *International Journal of Science Education, 29*(6), 685–710.

Eder, D., Evans, C. C., & Parker, S. (1995). *School talk: Gender and adolescent culture*. New Brunswick, NJ: Rutgers University Press.

Edwards, D., & Mercer, N. (1987). *Common knowledge: The development of understanding in the classroom*. London: Routledge.

Eldredge, N. (2009). To teach science, tell stories. *Issues in Science & Technology, 25*(4), 81–84.

Falk, J. H., & Dierking, L. D. (1992). *The museum experience*. Washington, DC: Whaleback.

Falk, J. H., & Dierking, L. D. (2000). *Learning from museums: Visitor experiences and the making of meaning*. Walnut Creek, CA: AltaMira Press.

Firestone, W. A. (1993). Alternative arguments for generalizing from data as applied to qualitative research. *Educational Researcher, 22*(4), 16–23.

Galton, M. (1998). *Inside the primary classroom: 20 years on*. London: Routledge.

Griffin, J. M., & Symington, D. (1997). Moving from task-oriented to learning-oriented strategies on school excursions to museums. *Science Education, 81*(6), 763–779.

Hargreaves, L. J. (2005). *Attributes of meaningful field trip experiences* (Unpublished Master's thesis). Simon Fraser University, Vancouver, Canada.

Independent Schools Council. (2010). *Teaching staff & teacher/pupil ratio*. Retrieved December 10, 2010, from http://www.isc.co.uk/publication_8_0_0_11_781.htm

King, H. (2009). *Supporting natural history enquiry in an informal setting: A study of museum explainer practice* (Unpublished doctoral dissertation). King's College London, London, UK.

Lemke, J. L. (1990). *Talking science: Language, learning and values*. Norwood, NJ: Alex.

Lemke, J. L. (1998). *Teaching all the languages of science: Words, symbols, images, and actions*. Retrieved July 12, 2010, from http://academic.brooklyn.cuny.edu/education/jlemke/papers/barcelon.htm

Lord, P., & Jones, M. (2006). *Pupils' experience and perspectives of the National Curriculum and assessment: A review*. London: Qualifications and Curriculum Authority (QCA).

Mercer, N., Dawes, L., Wegerif, R., & Sams, C. (2004). Reasoning as a scientist: Ways of helping children to use language to learn science. *British Educational Research Journal, 30*(3), 359–377.

Mercer, N., & Littleton, K. (2007). *Dialogue and the development of children's thinking: A socio-cultural approach*. London: Routledge.

Millar, R., & Osborne, J. (1998). *Beyond 2000: Science education for the future*. London: King's College London.

Morgan, S. C., Hamilton, S. L., Bentley, M., & Myrie, S. (2009). Environmental education in botanic gardens: Exploring Brooklyn Botanic Garden's Project Green Reach. *Journal of Environmental Education, 40*(4), 35–52.

Nagel, N. G. (1996). *Learning through real-world problem solving: The power of integrative teaching*. Thousand Oaks, CA: Sage.

Orion, N., & Hofstein, A. (1994). Factors that influence learning during a scientific field trip in a natural environment. *Journal of Research in Science Teaching, 31*(10), 1097–1119.

Osborne, J., Erduran, S., & Simmon, S. (2004). Enhancing the quality of argumentation in school science. *Journal of Research in Science Teaching, 41*(10), 994–1020.

Paris, S. G., & Turner, J. C. (1994). Situated motivation. In P. R. Pintrich, D. E. Brown, & C. E. Weinstein (Eds.), *Student motivation, cognition, and learning* (pp. 213–237). Hillsdale, NJ: Lawrence Erlbaum.

Paris, S. G., Yambor, K. M., & Packard, B. (1998). Hands-on biology: A museum-school-university partnership for enhancing students' interest and learning in science. *The Elementary School Journal, 98*(3), 267–288.

Rickinson, M., Lundholm, C., & Hopwood, N. (2009). *Environmental learning: Insights from research into student experience*. Dordrecht: Springer.

Ryan, A. M., & Patrick, H. (2001). The classroom social environment and changes in adolescents' motivation and engagement during middle school. *American Educational Research Journal, 38*(2), 437–460.

Sanders, D. (2004). *Botanic gardens: 'Walled, stranded arks' or environments for learning?* (Unpublished doctoral dissertation). University of Sussex, Brighton, UK.

Scott, P. H., Mortimer, E. F., & Aguiar, O. G. (2006). The tension between authoritative and dia-logic discourse: A fundamental characteristic of meaning making interactions in high school science lessons. *Science Education, 90*(4), 605–631.

Tal, T., & Morag, O. (2007). School visits to natural history museums: Teaching or enriching? *Journal of Research in Science Teaching, 44*(5), 747–769.

Tran, L. U. (2004). *Teaching science in museums* (Unpublished doctoral dissertation). North Carolina State University, Raleigh, NC., USA.

van Zee, E. H., & Minstrell, J. (1997). Using questioning to guide student thinking. *The Journal of the Learning Sciences, 6*(2), 227–269.

Wandersee, J. H. (1986). Plants or animals: Which do junior high school students prefer to study? *Journal of Research in Science Teaching, 23*(5), 415–426.

Wellington, J., & Osborne, J. (2001). *Language and literacy in science education*. Buckingham: Open University Press.

Zubrowski, B. (2009). *Exploration and meaning making in the learning of science*. Dordrecht: Springer.

Chapter 6
The Pedagogical Behaviours of Botanic Garden Educators

This chapter addresses the research question 'What pedagogical behaviours could be observed during the BGEs' guided visits in terms of supporting and enriching the students' learning?' As there is scant literature regarding the teaching behaviours of informal educators, the observation data collected for this study, including transcribed video data and field notes, were analysed to identify the pedagogical behaviours of the participating BGEs. According to their forms and functions, three categories of such behaviours were identified, these being class management (Sect. 6.1), pedagogical moves (Sect. 6.2) and teaching narratives (Sect. 6.3). Subsequently, a framework of the BGEs' pedagogical behaviours is developed, and the relationships among the three categories are discussed in Sect. 6.4.

6.1 Class Management

Gaining and retaining the attention of students is probably the most challenging part of teaching. Moreover, it is critical to positive educational outcomes that educators can maintain 'a potential balance between the targets of discipline which aims to control behaviour and discipline which aims to promote study' (Burke, 2007, p. 176). In other words, the ability of educators to organize the class and manage the behaviour of their students is a prerequisite for effective instruction. Drawing on the analysis of the observational data, this section presents the participating BGEs' practices regarding their class management, which covers organizing the group (Sect. 6.1.1), management talk (Sect. 6.1.2) and collaboration with schoolteachers (Sect. 6.1.3).

© Springer Science+Business Media Singapore 2015
J. Zhai, *Teaching Science in Out-of-School Settings*,
DOI 10.1007/978-981-287-591-4_6

6.1.1 Organizing the Group

Much research on school trips to informal settings has found that the educators' intervention, especially providing opportunities for group work and supporting active student participation, has a positive impact on student learning experiences (Stavrova & Urhahne, 2010). For the school trips to botanic gardens observed for this research, the participating BGEs were responsible for promoting student engagement and participation through effective group management, which involved the following strategies: grouping the visiting schoolchildren, assembling the group and nominating a speaker.

6.1.1.1 Grouping

According to the field observations and interview data, at the beginning of the guided visits, the BGEs nearly always divided the students into several small groups, which they explained had the purpose of providing each child optimal opportunity to interact with the BGEs and their peers. In gardens A and B, where the BGEs lacked support from education volunteers, the schoolteachers were invited to lead some teaching sessions. For example, Mark invariably divided the students into two groups (one led by him and the other led by a schoolteacher) to work on the same learning activity, simultaneously.

Grouping sometimes was determined by the nature of the learning activities or the availability of teaching resources. For instance, when students were learning about plant adaptation, Simon asked them to work in pairs exploring the glasshouse so as to find out the answers to the questions listed on their worksheets. Moreover, for pond dipping, Mark required them work in pairs or small groups, according to the number of trays placed near the edge of the pond. Once the students had finished data collection from dipping into the pond, they were directed to work in groups to observe and identify the pond insects in their tray, using the clues and pictures on the identification key sheet.

It emerged that dividing the students into groups made it more practical for the BGEs to monitor, support and assess individual student's learning. Further, this strategy gave the children more opportunities to communicate with the adults and ask questions that came from their visiting experience.

6.1.1.2 Assembling the Group

One key difference between learning in school settings and that in botanic gardens is that for the former most of the activities take place within a classroom, whereas for the latter the students need to move between places of interest. That is, the learning activities during the visits have to be carried out in different areas of the garden, and hence, the BGEs have to walk the students from one learning station to another. In order to draw students' attention to the instructions, two strategies were used when the BGEs assembled the groups, either before or after they had channelled them to a different learning site. First, there was the non-verbal and move only strategy, where on arriving at a new learning site, the BGE stopped and waited for the students to be standing around him/her, which was employed when the students were in a small-sized group and were behaving themselves. By contrast, the second strategy used was that employed when the group was relatively large in size, and the students were easily distracted by their surroundings, especially the public visitors, the plants and the animals in the gardens. In these situations, the BGEs would call on the students to line up or stand in a semicircle and remind them to pay attention to the instructions, with phrases such as 'hurry up', 'come over here' and 'listen', being frequently heard during this process. However, it was observed that sometimes the BGEs started teaching without assembling all the students together, and as a result some of them would miss part of the instructions or explanations. In this regard, as reported in Chap. 5, one student responded to the question about Mark's teaching on his feedback sheet, complaining that he started talking without having all the children present at the learning stations.

6.1.1.3   Nominating a Speaker

Nominating a speaker refers to a BGE giving students the authorization to answer questions, make comments or express opinions during the whole class/group instruction. On noticing a student who had risen a hand indicating they wanted to speak, the BGEs were observed using three forms of authorization: (1) pointing at the student who wishes to speak, (2) saying 'yes/yeah' and nodding their head and (3) saying 'yes/yeah' and pointing at that student. By nominating specific individuals using these methods, the BGEs were ensuring that there was only one speaker at a time so that everyone else would be able to listen to that student's contribution (Mercer, 1995). However, nominating a speaker did not necessarily have to involve there being a student who had actively put up a hand to speak, for instance, a BGE would frequently nominate those not actively involved, especially when they were invited to answer a question. For example, when Simon was teaching the BS School students 'plant and grow' inside the classroom, he often invited those children who were not actively engaged in participating to do so (see the following excerpt).

Simon: The petals have got a very special job. Why do you think it has these petals? They look good; they might smell good, what for? Yes. [points at S1 who raised his hand]
S10:   They can make the food.
Simon: It's not to make the food. They make their food in their leaves or in the stem if it's green. **Let's ask that boy sitting in the corner who keeps quiet all the time**. [points at S18]

S18:   It could be for the bee.
Simon: It could be for the bee, that's right. Yes [points at S3 who raised his hand].
S3:   It attracts the butterfly.
S9:   Wasps. [speaks out without asking to]

After asking a student to stick some red petal-shaped felt pieces on a flocked backing board so as to demonstrate how petals grow on flowers, Simon challenges the group about the function of petals. The answer from S10 is not the response that Simon has anticipated and so he then invites another student to respond. This time, however, he nominates S18, who has been keeping quiet since the start of the workshop, to share his opinion with the class. As the following sequences from S3 and S9 demonstrate, although S18's answer does not explain the function of petals very thoroughly, it provides the other students with a clue that petals can attract pollinators. The reason why S18 sat quietly and did not actively participate in the learning activities was not clear. It could have been because the student had less confidence regarding speaking out in class, or perhaps he was distracted by the novel environment of the botanic garden and, hence, could not concentrate on the lesson. In this regard, by nominating such students who were not fully participating in the activities, each was being encouraged to be a part of the classroom community by the BGEs.

6.1.2 Managing and Monitoring Student Behaviour

In Brophy and McCaslin's (1992) research on teachers' perceptions of classroom management strategies, many participants reported that they prefer neutral or supportive interventions over negative actions, but accepted that control-oriented strategies were appropriate for problem students. In their interviews, the BGEs expressed the opinion that managing and monitoring student behaviour in outdoor settings was more challenging than in school classrooms, and in this subsection the strategies they used to deal with disciplinary and behavioural issues are discussed.

As botanic gardens are open to the public, the visiting schoolchildren could have disturbed other visitors (e.g. by noise or by blocking the pathway when they are doing activities). Moreover, in a botanic garden medicinal plants are grown, which can be poisonous, and the animals such as peacocks that live in the garden could hurt the students if they felt threatened. Therefore, the BGEs have a responsibility to make students understand what is acceptable and what is not during the visit. In particular, the health and safety issue is a primary concern for them when leading a visiting school group in a garden. The following example shows how Mark reminded his students about the rules and regulations at the beginning of the visit, especially safety matters:

Mark: … The first rule in the garden is please stay with adults you are with. If you
         are walking around the garden please make sure you don't pick anything
         from the plants because these plants need their leaves, their stems, their fruits,

their roots, all that stuff, more than we do. Plus, some plants here are
deadly poisonous and some plants are super spiky. I don't want anyone to
be hurt or poisoned.

Apart from the health and safety issues, the students could have been distracted
by the botanic garden environment, due to the public visitors, the animals living in
the garden and other school groups. For example, when Mark was organizing a
whole class discussion about the habitats that human beings live in, some students
were still doing their observational drawings that they should have already finished.
Noticing those students were not focused on the discussion, Mark asked them to
stop sketching. Another example of getting the students back on task happened during Simon's workshop when the students were exploring the glasshouse to find
answers for their worksheets, and he noticed that some of them were just sitting on
the bench with their worksheets on the floor. So he asked them whether they had
found the answers to all the questions and urged them to stand up and keep exploring the glasshouse.

6.1.3   Collaboration with Schoolteachers

Research has shown that assistance from accompanying schoolteachers can help
ensure successful school trips to informal settings (DeWitt, 2007; Griffin, 1998;
Hargreaves, 2005), and accordingly a collaborative partnership between the BGEs
and the schoolteachers is essential for providing the students with a better learning
experience during the visit than it would be otherwise. Through analysing the observation data, there were two themes that emerged regarding BGE-teacher collaboration: 'promoting teacher involvement' and 'building/enhancing partnership'.

During the visit, most of the BGEs would enlist the help of the schoolteachers to
perform such tasks as dividing the students into groups, assembling the groups for
instruction and handing out learning materials and equipment, all which helped to
make BGEs' teaching more efficient. As well as asking them to assist with these
tasks, in Mark's case, he encouraged them to participate in hands-on, exploratory
activities with their students, which reflected good practice as suggested by Griffin
(1998). In their interviews, most of the BGEs agreed that it is important to encourage the schoolteachers to participate actively in the teaching session; however, in
practice this rarely occurred.

Communication between the BGEs and the schoolteachers during the visit can
be a good opportunity for them to build a rapport (Vergou, 2010), but in the current
research BGE-teacher communication to this end was observed only once. On this
occasion, when the SB School students were exploring the tropical glasshouse by
themselves, Simon was chatting with their schoolteacher about each other's teaching
experience in school and outdoor settings. Subsequently, the topic moved on to
the forest school project that the school was participating in and finally they were
discussing and planning another visit for the autumn term.

6.1.4   Summary

In the above subsections the strategies that the BGEs employed to manage their guided visits have been presented, and Table 6.1 sets out the detailed pedagogical functions of these strategies. It was observed that all the BGEs tended to group the students randomly, rather than ask for the schoolteachers' suggestions or the students' preferences. With respect to this, extant research into children's friendships and learning in primary schools has indicated that using friendship as a basis for classroom grouping for cognitive tasks can help promote their learning (Baines, Blatchford, & Kutnick, 2009; Kutnick & Kington, 2005).

Table 6.1 Categories of the observed BGE strategies in managing the guided visits

Class management	Description	Purpose	Conditions
Organizing the group			
Grouping the visiting students	BGE divides the whole class/leading group into small groups/pairs	To maintain high BGE-student ratio; maximize each student's participation in activities	At the very beginning of the workshop; when activities require small group/pair work
Assembling the group	BGE gets students together	To make students focus attention for instruction	When the BGE leads students from one learning station to another; when the students arrive at a new learning station
Nominating a speaker	BGE nominates a student in the group to speak	To guarantee only one speaker at a time so that everyone can hear that student's contribution; improve classroom discourse quality	Whenever the BGE notices a bidder; when some students are off-task
Managing student behaviour	BGE manages health and safety issues and student misbehaviour	To minimize potential danger; to get students focused on tasks	At the beginning of the visit; when student misbehaviour is spotted
Collaboration with schoolteachers			
Promoting teacher involvement	BGE encourages the schoolteachers to take part in student activities	To make students aware of the importance of the visit for their learning	During the student activities
Building/enhancing partnership	BGE discusses/chats with the schoolteachers about the preparation/ follow-up issues	To develop a better partnership for future collaborations	At the beginning and end of the visit

The strategies for managing and monitoring student behaviour have the purpose of securing a safe and well-ordered learning environment that engenders enthusiastic participation by the visiting schoolchildren. This can be best achieved by the BGE and the schoolteacher agreeing what the ground rules are beforehand. This was obviously not the case with Mark and a particular schoolteacher on one occasion, when the latter kept cutting across him whilst he was talking which I could see made him annoyed. Everton and Weinstein (2006) warn that class management strategies that depend on setting rules and enforcing student discipline can negatively influence the classroom atmosphere. Moreover, a substantial amount of research has found that inappropriate classroom management, in practice, is caused by inadequate teacher training towards establishing positive and productive classroom environments (e.g. Emmer & Stough, 2001; Evertson & Emmer, 1982).

The third strategy for class management, namely, collaboration with schoolteachers, was not widely observed in the participating BGEs' practices. In an investigation on collaborative relationships between botanic gardens and local schools in England, Vergou (2010) found that these were influenced by three factors, including (a) the inclusion of the National Curriculum in garden's learning program, (b) the inherent integration of the visit in the school work planning and (c) the continuity and trust, knowing what the collaboration can offer. The first two issues can be solved when the BGEs and the schoolteachers are planning and preparing the visit, but the last issue can only be handled during and after the visit. Therefore, the BGEs should be encouraged to get the schoolteachers on board with the purpose of developing a continuous collaborative partnership, rather than a one off encounter.

6.2 Pedagogical Moves

Pedagogical moves refers to the BGEs' physical actions to support and enrich the visiting schoolchildren' learning, and these are not simply combined with their educational talk to assist understanding, but they themselves can contain educational meanings. Next, based on the observed teaching practices of the participating BGEs, the meanings of these pedagogical moves are discussed according to three categories: 'pointing at plants', 'demonstration' and 'the use of models'.

6.2.1 Pointing at Plants

6.2.1.1 Plant Pointing as Referencing

Plant pointing as referencing was used when a BGE indicated a particular plant, when talking, so as to help students identify which one was under discussion. For example, when Mark was taking the FP School students around the glasshouse in order to compare the living habitats of succulent and carnivorous plants, he pointed at a *Pachypodium* to show them its location:

S3:   If I had to fall on one of these cactus plants which one would hurt most [points at the cactus plants in the glasshouse]?

Mark: If I had to fall on one of these cactus plants which one would hurt most? Hmmm. I think the Madagascan one which is really tall and super spiky.

S3:   Which one?

Mark: [points at the *Pachypodium* next to the wall] The big one in the pot, just on the second shelf in the middle. That one would be quite painful because it's got quite a lot of spiky leaves.

From the above excerpt, it is apparent that Mark points at the *Pachypodium* when S3 is unsure about which is the plant that he has mentioned, this being the tall and super spiky one from Madagascar. By pointing at the plant, he assists the students in navigating their way to pinpointing the location of the *Pachypodium* in the glasshouse. Another example of this movement is when he shows the students the pitcher plants, where in order to guarantee each student is looking at them before going on to explain their predation process, he points at all the different ones located in front of the students and verbally informs them which direction their attention should take:

Mark: [points at the pitcher plants in front of the students] This one and some next to you in front there, there're some reddish ones, there're some in front of your hand, yeah, right in front of you and there're some right here and some here. Hopefully you are able to see one close up....

Deictic gesture, a term introduced by McNeill (1992) to describe the movement of finger pointing, refers to both concrete and unobservable objects, such as processes. During the observed guided visits, the deictic utterances, such as 'here', 'there', 'this' or 'that', were widely used by the BGEs when referring to an object, and because these utterances often held ambiguous meanings for the listeners, they would use finger pointing to clarify the object to which they were referring.

6.2.1.2 Plant Pointing for Explaining and Describing

The BGEs pointed at plants, not only as a form of referencing but also for the purposes of explaining and describing. According to McNeill (1992) and Crowder (1996), thought is mediated by both language and gesture, and thus, gestures that are associated with explanations and descriptions can assist in conveying an idea clearly. For example, after showing where the pitcher plants are to be found, Mark catches hold of one of the ones next to him and starts to talk about it. He points out the different parts of that particular plant whilst he is describing and explaining their structures and functions, as shown below.

Mark: [points at the pitcher tube] This is basically a leaf which is like a deep dark well full of some liquid. [points at the operculum which looks like a lid for the pitcher plant] This is the little umbrella which keeps the rain out [points at the peristome]. And at the back there's some very sweet stuff. It's quite shiny. It is nectar, a bit like honey.

However, in practice, plant pointing for referencing or explaining did not always occur. Instead, sometimes the BGEs talked about certain plants without drawing the students' attention to which plant or which part of a plant was under discussion, with the consequence that often the students could only get a vague grasp regarding the object they were actually talking about.

6.2.2 Demonstration

Demonstrations were regularly observed in the BGEs' guided visits, especially during the whole group instruction. Through demonstrations, a BGE can focus students' attention on the object that she/he is operating and engaging them in listening, observing and thinking. In this study, the BGEs' demonstrations were divided into 'technology-assisted' and 'task-oriented' demonstrations, according to their forms and functions.

6.2.2.1 Technology-Assisted Demonstration

Rather than only showing students authentic objects, such as plants or insects that live in the garden, the BGEs can use digital technology equipment to help them make sense of the biological, geographical and environmental knowledge. During the observed guided visits, technologies, such as CD-ROM, microscopes, LCD projectors and computers, were used as additional supportive tools to present the topics.

The use of technical equipment can extend students' observations to include evidence of the kind that would otherwise be imperceptible to them (Harlen, 2000). For example, Mark used a specially designed microscope that was connected to an LCD projector, to demonstrate pond life to visiting students. First of all, he carried out an experiment by drawing a water sample from the bottom of the fish tank and then dropped several water drops onto a microscope slide, before placing it under the object lens. Next, he magnified the sample water by 650 times and some tiny moving creatures appeared on the screen. Finally, he switched to another lens and zoomed it in to 1,500 times magnification, during which time some large quick moving creatures showed up on the screen. To give students a clear concept about the scale of the image they saw on the screen, Mark explains that the entire screen width represents one millimetre in real terms:

Mark: Some of you were asking about the microscope—what does it do? This is microscope which allows very small things appear much, much bigger [points at the microscope next to the fish tank]. What I'll do is to put some of the stuff from the bottom of this fish tank underneath the microscope to show you [draw water sample by using a dropper pipette, drops the sample on a microscope slide, places the slide under the object lens]…What you are seeing right now is magnified by 650 times, if I zoom it in, it is magnified by 1,500 times [changes the object lens].… Well, to give you an idea what is happening here [points at the screen]. If you think of a ruler. The smallest division of a ruler is a millimetre. That width of that screen is one millimetre, pretty much exactly. The screen is a metre and half wide and the magnification is 1,500 times, so that's what is happening inside one millimetre. So that's the home for creatures we normally can't see.

Some phenomena or processes, such as animal behaviours, are difficult for students to observe during a visit. For example, the nesting behaviour of blue tits is not easy to see as it takes place over a protracted period of time, so when he was teaching the NP School students about habitats, Mark showed them a video clip which recorded the process of a blue tit making a nest in a bird-box. This video clip that had been recorded on a mini-camera installed inside the bird-box showed the whole nest-building procedure. Although watching such a video clip and other things similar offer vicarious experiences, they can help to facilitate students' conceptual understanding of a complex scientific phenomenon or process.

PowerPoint software was a useful digital technology used for demonstrating ecological science in the BGE-guided visits at gardens B and C. In Garden B, there were visits designed for the visiting secondary school groups, which focused on the tropical rainforest, one part of which being the BGE's PowerPoint format presentations based in the classroom. Through these presentations, Debbie, the BGE who was in charge of visiting secondary groups at Garden B, showed many diagrams and pictures to facilitate students' sense making about the ecological environment in the tropical rainforest and the loss of biodiversity caused by deforestation. Julia, the BGE at Garden C, instructed students about seeds with the assistance of a PowerPoint presentation, which showed magnified images of different seeds on the screen to help students understand their structure and the germination process.

6.2.2.2   Task-Oriented Demonstration

Whilst technology-assisted demonstrations can be used to support BGEs' explanations of scientific ideas or processes, the purpose of a task-oriented demonstration is to prepare the students for a hands-on activity. For some activities, such as pond dipping or planting, the BGEs usually demonstrated how to use the relevant equipment before the students started their own work. In the example below, before the students are allowed to go pond dipping, Mark demonstrates how to use a scoop net in the pond to collect water insects. They are gazing at the net in the BGE's hand and listening to the instruction about how and where they should scoop it so as to collect a good range of samples from the pond:

Mark: With your net [picks up a net from the floor], you need to sweep around in the water, backward and forward, maybe scraping the walls of the pond itself [sweeps the net in the water], maybe do some scooping motions like that among plants under the water [scoops the plants in the pond]. We have got the algae called blanket weed, which you can put back in [takes out the blanket weed from the net and throws it into the pond]. You need to turn your net upside down and put the samples you collect into the tray [turns over the net and drops the samples into the tray below].

In addition to task-oriented demonstrations explicating the correct procedures for hands-on exploratory activities, this pedagogical move can reduce the novelty concerning any equipment or activities that the students are unfamiliar with.

6.2.3 The Use of Models

Models are key tools in science learning, especially in aiding understanding of complex abstract concepts and phenomena (Coll, France, & Taylor, 2005; Gilbert, Boulter, & Elmer, 2000; Harrison & Treagust, 1996). Although most parts of the BGEs' teaching focused on addressing concrete and factual knowledge of ecology, on some occasions models were observed being used to communicate complicated concepts.

6.2.3.1 The Use of Physical Models

Physical models are scaled miniatures or enlarged working models used to help explain complex ideas to students, and the BGEs had designed different types not only for demonstration but also for students' hands-on activities. During the observed guided visits, two kinds of physical models were frequently used when the BGEs were instructing the students about the structure of the plants.

Simon used what he called fuzzy felts[1] to facilitate the SB School students recalling from their memories the structure of a plant and the external conditions that support growing that they had learned in school. During the lesson, he invited the students to construct a picture of a growing plant by putting felts of different shapes, which represented the root, stem, leaf and petal, on the board, and this activity jogged the students' memories regarding their previous knowledge about different parts of a plant and their scientific names.

---

[1] Fuzzy Felt ® is a fabric toy for young children. The toy consists of a flocked backing board onto which a number of felt shapes are placed to create different pictures, being attached rather like Velcro as used in clothing.

Similarly, Julia had designed some dissectible plastic flowers for the school groups who were scheduled to study flower structure during the visit, and the students were encouraged to work in pairs to construct the flower model when she had finished the instruction regarding their structure.

6.2.3.2   The Use of Metaphorical Models

Metaphorical modelling refers to the hand or arm movements that depict an imaginary object or concept. McNeill (1992) named this kind of body movements as a metaphoric gesture, which show pictorial content and abstract ideas at the same time, and Roth and Lawless (2002) argue that such gestures have a narrative character, whereby the images used can help explain abstract processes. The BGEs sometimes resorted to metaphoric gestures to assist their explanation of certain abstract phenomena or ideas. For example, when Mark was leading the FP School students in the carnivorous glasshouse, he introduced the sundew plant to the students, by first explaining the trick that it uses to prey on insects, that is, the secretions to attract and ensnare them. Then, he informed the students that 'What happens is if a fly lands on one of the bits of leaf and gets stuck to it', whilst stretching out his right hand index finger to touch the palm of his left hand (see frames 1–4 in Fig. 6.1). His finger and hand movements represent the process of a fly landing on a sundew leaf and becoming stuck because of the secreted sweet mucilage. Subsequently, as he continued, 'The leaf will slowly bend to sandwich that fly, literately, the fly sandwich', and slowly bent his left palm to symbolize this processes (see frames 5–8 in Fig. 6.1). The metaphorical modelling used here can act to mediate students' understanding of abstract subject matter. Moreover, with the movements being associated with language, this helped to create vivid images for the students to conceptualize the preying

Fig. 6.1 Metaphorical modelling of the predation process of a sundew

mechanism of sundews as if it had happened for real, hence further enhancing their understanding. In sum, with the assistance of metaphorical gestures, unobservable processes in real time or space can give form that supplements written and spoken language, thereby assisting comprehension of complex things and ideas (Roth & Lawless, 2002).

6.2.4 Summary

The BGEs' pedagogical moves presented above can not only capture the students' attention for listening to instructions, but they can also help shape their ideas about abstract ecological concepts and processes (see Table 6.2 for their detailed pedagogical functions).

Table 6.2 Categories of the observed BGE pedagogical moves

Pedagogical moves	Description	Purposes	Conditions
Plant pointing			
Plant pointing as referencing	BGE points at a plant to indicate the direction of it	To minimize the vagueness of the plant that the BGE refers to	When the BGE mentions a plant by using deictic utterances (e.g. this, that, here, there, etc.)
Plant pointing for explaining and describing	BGE points or holds a plant for explaining or describing	To guarantee each student can have a clear image of the plant that the BGE is going to talk about	When the BGE is going to give a detailed explanation about or a description of a particular plant
Demo			
Technology-assisted demonstration	BGE uses ICT to assist demonstration	To provide students with enlarged image of micro objects; to make unobservable phenomenon observable	When teaching abstract ideas; when some objects cannot be observed through the naked eye
Task-oriented demonstration	BGE illustrates how to conduct hands-on activities	To reduce the novelty issue; to maximize the efficiency of student activities	When the activity involves manipulating equipment
The use of model			
The use of a physical model	BGE uses physical models for instruction	To help students conceptualize abstract concepts; to improve motivation and engagement	When the subject matter is abstract or difficult for young children to understand
The use of metaphorical models	BGE uses metaphoric gestures to assist explanation	To assist students to make sense of abstract or non-observable subject matter	When addressing abstract and non-observable issues

The plant-pointing movements assisted the students in identifying the objects that a BGE was talking about on the spur of the moment. Regarding the demonstrations, these helped the students develop a better understanding of how an experiment could or should be carried out. Moreover, the ICT-assisted demonstrations, such as the PowerPoint presentations used by the BGEs, served in 'focusing attention on over-arching issues' and 'increasing salience of underlying abstract concepts' (Osborne & Hennessy, 2003, p. 4). The use of models was another strategy that the BGEs employed to support their students' learning. For example, the use of metaphorical models, such as that of Mark's sundew described above, is 'simplified representations of phenomena or ideas which take up an intermediate position between reality and a mental model' (Coll et al., 2005, p. 185).

6.3   Teaching Narratives

From the sociocultural perspective of learning, knowledge is co-constructed through participation in socially and culturally organized activities (Lave & Wenger, 1991; Wertsch, 1991). As a mediating tool of meaning making, language plays a significant role in the activities of teaching and learning (Edwards & Mercer, 1987; Vygotsky, 1987; Wells, 1999). Based on the Scott's (1998) framework of teacher narratives and the analysis of observation data collected for the current research, here, the strategies that the participating BGEs adopted to communicate ecological science to the students are grouped into 'developing the conceptual line', 'shaping the epistemological line', 'promoting shared understanding', 'checking student understanding' and 'explanatory talk'.

6.3.1   Developing the Conceptual Line

The central role of BGE talk is to develop the conceptual line so as to help facilitate student understanding. More specifically, developing the conceptual line, for as Scott (1998) suggested, contributes to the portrayal of constructing shared knowledge during the course of classroom interaction. Based on the observations of participating BGEs' teaching practices, two kinds of talk that served to develop the conceptual line were identified, namely, shaping ideas and uptake response.

6.3.1.1   Shaping Ideas

Shaping ideas refers to the BGEs' talk which aimed at supporting students' knowledge construction, and this kind of talk can be divided into introducing new terms and ideas and revoicing student ideas, which are discussed below in detail.

The BGE-guided visits should not only help to reinforce the subject knowledge that the students acquired in school classrooms, for new terms and ideas need to be introduced as well. Instances of introducing new terms or ideas in a directed way without prompting or scaffolding were commonly observed across the different BGEs' practices. However, new knowledge was communicated in other ways, such as when the BGE created a familiarized context to facilitate the students' sense making regarding new concepts. For example, Simon introduced the term photosynthesis by creating an interesting condition to support students' construction of such an abstract concept:

Simon:	Come here Brandon [S3 goes to the front of the classroom]. If I took Brandon outside and planted him in the ground.
S11:	It would not be nice.
Simon:	It would not be nice but because he's not very strong. Brandon, I'll give you a big stick. You wrap around it [Brandon holds a big bamboo stem]. If I did take Brandon outside and left him outside in the sunshine with the air and watered him. If we left Brandon for 6 weeks what would happen to him? All he's getting is sunshine, water and air. Yes [nominates S7].
S7:	He would die.
Simon:	He would die, that's right. He would die because what hasn't he had?
S10:	He hasn't had water.
Simon:	He has water. What hasn't he had?
S3:	Sunshine.
Simon:	He had sunshine.
S2:	He hasn't had any food.
Simon:	He hasn't had any food that's right. We cannot make food. Plants can make their own food. They make it from a gas in the air. Who has heard of the word carbon dioxide? That's the gas they use to make their own food with energy from the sun and with water as well. We can't do that we have to purchase our food.
S5:	We also breathe out carbon dioxide.
Simon:	We do. It's a bit complicated. We breathe in the air and we take oxygen from the air in our lungs and that burns our food and gives us energy. Plants do that as well. Plants don't walk around and they don't need too much energy. But luckily when they are making their own food from carbon dioxide and sunshine and water, they actually make oxygen as a waste product. We do breathe out carbon dioxide but plants need oxygen as well. They can use carbon dioxide to make their own food. It's called photosynthesis. That's a great word and you have got a couple of years to learn that.

The idea of leaving Brandon outside in the conditions that plants have for survival (air, soil, water and air) is interesting to the Year 3 students, and by providing a strange but funny context, Simon is able to get their engagement whilst he explains to them the challenging concept of photosynthesis.

Revoicing is a form of educator talk which modifies the content or the vocabulary of a student's response. O'Conner and Michaels (1996) suggest that it can be used to imbue contributions from learners with scientific language and phraseology, whereby the educator can provide not only conceptual but also linguistic scaffolding. The dialogue below took place when Julia was introducing students to the glasshouses that they are going to visit.

Julia: The next house after the Carnivores is the Arid Lands House. Do you know
           what arid means? There is a clue on that picture (a cactus).
S5:   It means desert.
Julia: Arid means dry and desert is not the only one dry place.

It is apparent that the student's response is influenced by Julia's clue, a picture of a cactus at the entrance of the arid glasshouse. In order to provide all the students in the group a proper meaning of 'arid', she corrects what S5 has said, further explaining that arid means dry and that deserts are just one kind of an arid environment.

6.3.1.2   Taking Students' Response into Account

Uptake of a student's response occurs when the BGE uses a student's answer for further discussion, and this is an important component of Nystrand's (1997) dialogic exchange in classroom interaction, as when employed it shows recognition of the value of student contribution. The following transcript was excerpted from the scenario when Mark was challenging the students' idea about the plants that human beings eat.

S6:   Sugar comes from a plant.
Mark: Yeah. Do you know which plants we get the sugar from?
S7:   Sugarcane.
Mark: Our sugar comes in two forms. Some of our sugar, mostly the brown sugar
           comes from sugarcane which is a type of grass as well [shows a dried
           sugarcane to the students]. But some of our sugar comes from a plant
           grown outside in this country is called?
S3:   Sugar beet.
Mark: Sugar beet, yeah, which I'll show you in a moment.

What Mark does here is elaborating S6's contribution 'sugar comes from plant' by asking a question based on it. The response from S7 is correct but does not fully cover the answer, so Mark tells the students that there is another plant that grows in England that produces sugar, and by so doing he is cueing the students to complete the answer: sugar beet. That is, the question 'Do you know which plants we get the sugar?' uptakes S6's response 'Sugar comes from plant' for further discussion in the class.

6.3.2 Shaping the Epistemological Line

The epistemological line plays a significant role in understanding the scientific viewpoint, for as Scott (1998) argued, 'learning the science way of knowing involves not only learning how to use the conceptual tool of science but also coming to appreciate the epistemological framing of those tools' (p. 57). There were very few examples observed in the BGEs' talk that contributed to the epistemological line, but nevertheless, some of them did present epistemological features to the visiting schoolchildren. In order to give some idea of the importance of developing students' epistemological line in informal settings, these few instances involving such features are considered next.

6.3.2.1 The Scheme of Binomial Nomenclature

During a workshop for the students from the FP School, Mark explains why garden plants are labelled in Latin. His brief explanation illustrates some epistemological features of botany, this being what plant taxonomy is and how plants are named:

Mark: Ok. We will probably notice all the plants displayed here in the garden are labelled with scientific names. This is the language of Latin. Do you study Latin at all?

Ss: No.

Mark: It's a dead language but in the garden especially botanic garden, it's a language which is still alive. [points at the labels next to the plants] This bed that is being planted to show Latin names are really easy to identify. The reason we have Latin names in the first place is because, again, if you are a doctor you say go to pick a handful of daisies. A lot of things can be called daisy. We know what daisy looks like. But there are some chamomiles over there which are daisy-like [points at the plants around the daisy]. You can go to another country and find something that is very much like daisy. It might not be the right thing. So the Latin name,… basically what I'm trying to say is there are a lot of things called daisy but there's only one thing called *Bellis perennis*, which is the Latin name for a daisy. This is a different type of clover [points at the clover in the bed]. *Trifolium* means three leaves: 'folium' is leaf and 'tri' is three. So these are Latin names which are easy to generally translate. The man called Linnaeus was a Swedish botanist. Two years ago was his 300 years birthday so it's quite a long time ago. He was the person devised naming things with two Latin names and that revolutionized the world of naming animals and plants.

Mark's talk starts with drawing the students' attention to the labels next to the plants. The notion of Latin as a dead language in modern society but still alive in the botanic gardens might provoke the students' curiosity about why this language is used to label plants. However, rather than asking the students to think about this question, he exemplifies the reason by showing them different daisy-like plants. Furthermore, the idea about binomial nomenclature is briefly introduced through the translation of *Trifolium*, the Latin name for clover. Finally, Mark explains that the method of using two Latin names to name plants and animals was devised by Carl Linnaeus, the Swedish botanist. In sum, here he has managed to introduce to those Year 5 students fundamental knowledge regarding botanic terms, including 'Latin language', 'binomial nomenclature' and 'Linnaean taxonomy'.

6.3.2.2   What Is Good Observational Drawing?

The study of ecological science requires careful observation and description, and drawing is an excellent way to describe an object as well as understanding it better, for to make an observational drawing, the observer has to 'move beyond simple, mental images of what he believes a particular plant looks like and instead concentrate on the unique identity of that specimen' (Dempsey & Betz, 2001, p. 271). During the visit in the botanic gardens, the students were given plenty of opportunities to observe and draw the plants or plant artefacts. However, the BGEs spent little or no time on students' drawing skills and on assessing their ability. As a result, many students were left in the dark about the quality of their drawings even though they had a strong desire to draw. The following example is an exception which shows how one BGE developed the students' understanding of what a good observational drawing should be.

In the arid glasshouse, Simon has assembled the group when most of children finished their drawings of a plant with thick juicy leaves. When all the students have sat on the floor and become quiet, he starts to review their observational drawings:

Simon: Somebody came to me and said 'I am not very good at drawing'. I looked at her picture and I could identify that picture from her drawing. That means it's a good observational drawing. You don't need to be a brilliant artist, but looking carefully is what it all about. [Simon takes out that student's worksheet and shows her drawing to the whole group] I can identify it, it's a money plant. It shows a lot of details. I can see the patterns. I can see the shape. So that's an important thing we can recognize your drawing we know what plant it is.

With the student's concern regarding the quality of her drawing, Simon raises the issue to the whole class and explains what a good observational drawing should comprise. He distinguishes two kinds of drawings: biological and art forms. When referring to the student's work, he introduces the criteria to assess a good observational drawing, that is, it should record the unique features of an object and thus be identifiable to audiences.

The reason for attributing the above example to shaping the epistemological line is because botanists have a long tradition of using observational drawing to describe and record plants. For example, Darwin's sketching of plants during his voyage around the world with the HMS Beagle made a great contribution to the discovery of plant evolution and the development of ideas for his seminal book *On the Origin of Species* (Kohn, 2008). Therefore, understanding how to do observational drawings is an essential skill required for those learning botany.

6.3.3 Promoting Shared Meaning

According to Scott's (1998) framework of teaching narrative and Edwards and Mercer's (1987) argument on common knowledge, BGEs' talk that promotes shared meaning can be divided into 'presenting ideas to the class', 'sharing the experience of individual students with the class', 'sharing the group findings with the class', 'repeating a student's idea to the class' and 'using the 'we' statement'.

6.3.3.1 Presenting Ideas to the Class

During the whole class instruction phases, such as informing students about a fact, making demonstrations or explaining a concept, BGEs can make ideas available to all students. In this study, presenting ideas to the whole class was the event most often observed in the BGEs' talk. For example, in Debbie's workshop for the students that focused on the Amazon Rainforest, environmental issues were presented through PowerPoint slides.

Debbie: Remember that in every rainforest the same size as the lawn outside can be cut down every 2 s. Every 2 s there's a much rainforest disappeared as the size of our lawn. We all know rainforest disappears for different reasons, such as: roads, dams, mines, cattle. However, this is probably the number one threat to rainforest at the moment—oil palm [showing the PowerPoint slide]. It is supposed to be in about 1/10 of our products in supermarkets, including: crisps, mayonnaise, biscuits, soups and all sort of different things. You might be buying and eating products that come from rainforest, because oil palms are tropical plants. It has to grow in tropical environment so the rainforest clearly makes its way for it. That's why it's a big threat to rainforest, particularly affecting the Amazon where you are learning about. As this one—soybean [showing the PowerPoint slide]—is particularly affecting Brazil in quite a big way right now, because again it's a rainforest plantation. Lots of different food is made from soybeans, for example, burgers. Again, you are buying, eating those foods and without realizing what they contain. So it's another threat to rainforest right now.

Heavy logging and agricultural clearance are the main threats to the Amazon Rainforest, and here, Debbie is communicating this environmental issue to the students through lecturing.

6.3.3.2 Sharing the Experience of Individual Students with the Class

Fauna and flora are attractive objects for young children, which may incite their curiosity. In the next excerpt, a student expresses her experience to the BGE (Mark) when she sees a ladybird wriggling around inside its cocoon:

S3:   I touched one of these ladybugs it stood straight up.
Mark: Did you?
S3:   When I touched it again it went down.
Mark: [raises his voice to make sure all the students in the group can hear] Ok.
          Someone just told me she touched one ladybird and it stood up and when
          she touched it again it went down. It's actually in a process of pupation so
          it's wriggling around inside its cocoon. That is what you were watching.
          Try not to touch them because they are so fragile. As you can imagine it's
          a very vulnerable time to be. It can't defend itself. Something can easily
          come along and squeeze and hurt it. It's in the stage of changing from what
          we saw in my hand to a ladybird. So it's going through a massive transfor-
          mation. It's wriggling around changing into its new skin. It's like getting
          changed in a changing room.

By sharing an individual student's experience, the BGE introduces the concept of pupation, a biological process, to the group, and as a result, he not only addresses that individual student's puzzlement but also helps the others to realize how a ladybird behaves during the pupation stage.

6.3.3.3 Sharing the Group Findings with the Class

The BGE-guided workshops included a lot of exploration tasks, during which the students were assigned to work in groups, and sometimes, the group findings were meaningful to the topics covered in the workshop. The excerpt below is an example of a BGE sharing group findings with the class. Julia, the BGE at Garden C, is guiding the students in learning the topic of habitat by pond dipping.

Julia: There's a group found some bloodworm. The reason it has colour, have you
          heard of something called haemoglobin?
Ss: No.
Julia: You have got that in your blood and they have got haemoglobin inside
          them. You use it to absorb oxygen. Ok? They use haemoglobin as well.
          They need it in these muddy places at the bottom. So you can go down a bit
          and rummage into the mud.

The bloodworm is the larva of a non-biting midge which lives under the water; however, only one group has found that species. Julia informs the class that there are some bloodworms in the pond and that one group has already found them. She further explains that the haemoglobin helps the bloodworm to absorb oxygen and survive in the mud.

6.3.3.4 Revoicing a Student's Idea to the Whole Class

Revoicing a student's idea is another perspective of promoting a shared meaning. Research on classroom discourse in different contexts suggests that an educator's repeat is more than duplicating the response, for it draws the attention of the whole class to the utterance and emphasizes the educational significance of that particular response (Edwards & Mercer, 1987; King, 2009; O'Connor & Michaels, 1996). The excerpt below has Julia explaining plant adaptation by telling the students about the plants that live in dry environments. Julia is leading the students from the UW School in the arid glasshouse, and she tells them that the plants displayed in that room have been collected from Australia and South America.

Julia: Plants have adaptations. Do you know what adaptation means?
S1:   It means with the change of the weather, the plants develop to fit the local environment, hmm, which is good for it.
Julia: Yeah. Overtime plants have changed, plants and animals and people they are changing over time. As the plants in the dry places, change they will come up with different strategies that make it easy for them to live where they do. [Then Julia guides the students to observe a eucalyptus tree and summarizes its adaptation strategies]

6.3.3.5 Using the 'We' Statement

Throughout the BGE-guided workshops observed in this study, the BGEs frequently used the 'we' statement to express what has been done and what to do next. The use of the 'we' statements can give students the feeling that they are members of the 'classroom community', which, according to Edwards and Mercer (1987), emphasizes their central role.

For example, when Debbie was teaching the students from the WM School about the Amazon Rainforests, she used 'we' statements to create a shared experience. The excerpt below describes a scenario where she challenges the students to offer the reason why bananas have giant leaves:

Debbie: Big leaves, why? Why do they have big leaves?
S11:   To catch more sunlight.
Debbie: Simple as that photosynthesis. These are low growing plants. They stay down at the shrub layer. Do you remember that we said that it's really dark down on the forest floor? The shrub layer might get more percentage of the sunlight on the forest floor by having big leaves.

The question 'Do you remember that we said that it is really dark down on the forest floor?' reminds the students that they discussed the layers of a rainforest at the very beginning of the workshop. Therefore, the idea of 'the bottom of a rainforest is dark' is shared information for the whole class.

6.3.4   Checking Student Understanding

Research on teaching and learning in science classrooms has found that checking the meanings and understandings that students develop is an effective strategy that the teachers can use to support meaning making (Scott, 1998). In the current research, the BGEs' approaches to this can be divided into 'checking previous student knowledge' and 'checking consensus'.

6.3.4.1   Checking Previous Student Knowledge

The thermometer is a useful instrument that students can use during their visit to a botanic garden, for example, to measure the temperatures of different glasshouses. However, young children might not have a clear idea what a thermometer is for. In the excerpt below, Simon introduces the thermometer to the SB School students because they need to complete a worksheet later on, when they have explored the glasshouses:

Simon: This is a thermometer [shows the students a thermometer]. What do we measure with this thermometer? Yes please [nodding at S1].
S1:   Umm….
Simon: Would you tell me if you remember the word?
S1:   To measure how hot it is.
Simon: To measure how hot it is. What's the proper science word for that—how hot or how cold something is?
S9:   The temperature.
Simon: The temperature, well done. It measures temperature.

The purpose for Simon to elicit students' thinking by giving them clues is to check their previous knowledge about thermometers. It is easy for the children to reply that a thermometer is used to measure the range of hot or cold, but the answer he expects to hear is 'temperature', a scientific term, rather than 'how hot or how cold'. Through checking students' previous knowledge, the BGEs can identify the current level of the students' knowledge and link what they already know to the garden visit.

6.3.4.2   Checking Consensus

The students may have different points of view about a certain concept and in order to ensure that there is consensus around the correct meaning, the BGE can open up a dialogue with one student to the whole group. The following transcribed interaction during a BGE-guided visit shows an example of how Simon checked whether there was consensus regarding the pattern of leaves on a plant.

Simon is reviewing the NP School students' previous knowledge on plant structure. Once a student has worked out the name of each part of a plant, he asks him to
place felt (stem, leaf, root, petal shape, etc.) on the flocked backing board.

Simon: Which parts of plants usually grow from inside the stem?
S15:   The leaf.
Simon: The leaves. Come here Miss Leaf. Okay. Two leaves today [gives S15 two
         leaf-shaped felts]. You could attach it to the stem. What is your name?
S15:   Zeenat [places the leaf shaded felts in a symmetrical pattern on the board].
Simon: Right. Zeenat has put these two leaves around the stem. Is that right?
S9:   No.
Simon: It's not. How do you know that? You show me. [Simon takes off the
         leaf-shaped felts and gives them to S9]
         [S9 puts the drip tip side of leaf onto the stem]
Ss: Noooo! [the students are laughing]
Simon: Absolutely not. Could you put them on? [nominates S3]
         [S3 puts the leaves in an alternate pattern on the board]
Simon: Both of those are right [places leaves in both symmetrical and alternate
         patterns]. Leaves can grow opposite each other like that. They can grow
         in step. The big word for that is a Year 5 word. Who has learned the Year
         5 word? Right, that is called alternate. When there is a step like that: one
         there, one there [shows what alternate pattern looks like].

Actually S15's layout of leaves is correct, but Simon still seeks others' opinions
by asking 'Is that right?' It seems that S9 does not agree with S15 and places the
leaf-shaped felts differently. As S9 finishes his work, all the students exclaim 'no'
and laugh at him, making it clearly apparent that what S9 has done is not accepted
by the class, so therefore, Simon nominates another student (S3) to rearrange it in
the way he thinks is correct. Subsequently, whilst putting S3 and S15's patterns of
leaves on the stem, Simon comments that both of them are right and introduces the
scientific terms for those patterns, namely, symmetrical and alternate.

Checking consensus is an important element of teacher talk, which has been
identified in other research. For example, O'Conner and Michaels (1996) use the
term 'align' to describe this form of teacher talk and pointed out that it credits a
particular role to students, such as predictor, theorizer or hypothesizer. In the above
example, Simon gave the students the role of 'evaluator'. King (2009) highlights the
pedagogical value of this kind of talk, characterizing it as 'as a way of guiding students
with an opportunity to rehearse meanings under the guise of a particular role' (p. 172).

6.3.5   Explanatory Talk

The explanation of scientific ideas requires 'assigning, developing, or expanding
meaning; offering a justification; providing a description; or giving a causal account'
(Harlow & Jones, 2004, p. 546). However, the act of explaining science is mainly

anecdotal, lacking any systematic or thought-out basis (Ogborn, Kress, Martins, & McGillicuddy, 1996). Apart from lecture-type explanations, the observed BGEs used a variety of strategies to communicate ecological science to the visiting school-children, including 'explanation by comparing', 'the use of analogy and metaphor' and 'scientific storytelling'.

6.3.5.1   Explanation by Comparing

Comparison is one fundamental approach of meaning making in the learning of science, and the BGEs used it to explain scientific ecological facts. Moreover, the BGE's employment of this strategy can be divided into two groups according to the level of complexity: 'superficial comparison' and 'structure mapping'.

Young children usually refer to a damselfly as a dragonfly, which is not correct from a scientific perspective: a damselfly belongs to the suborder Zygoptera, whilst true dragonfly belongs to *Anisoptera*. The following excerpt is an example how a BGE explains the subtle differences between a damselfly and a true dragonfly to the SF school students:

Julia: [holds two pictures in her hand] Can you all see these pictures?

Ss: Yeah.

Julia: These are both dragonflies. Ok? This is a type of dragonfly called a damsel-fly [points at the picture of the damselfly]. In the dragonfly group, there are damselflies and there are other animals more like this, called true dragonflies. Look carefully at these photos and look for differences. Don't worry about colour; don't worry about what they are sitting on. Look at differences in shape and how they put their wings together. Tell me what you can see.

S1:   [Many children raise their hands] That one is skinny and longer. [points at the damselfly]

Julia: Yeah, that's one of the differences. So the damselfly has a long thin body and the true dragonfly has got a much chunkier stouter body [points at the picture of dragonfly]. Well done. What can you see?

S2:   Its wings are a lot thinner.

Julia: There is a difference with the wings but it's not so much thin. If you look at how they are holding their wings, can you see this damselfly is holding its wings along its body? And the dragonfly is sticking them out like that [stretches out arms and simulates the dragonfly's wings as shown in the picture]. That's pretty well true there are a few damselflies that stick them out a little bit. Basically dragonflies stretch them out; damselflies hold them along the body. There's one another major difference you can see in these photos. What can you see?

S2:   The eyes are different.

Julia: It is about the eyes. You look at these eyes [points at the picture of damsel-fly]. Can you see the gap between the eyes?

Ss: Yeah.
Julia: On this true dragonfly can you see the eyes touch? [points at the picture of
          true dragonfly]
Ss: Yeah.
Julia: Okay. There is actually one dragonfly in this country where the eyes don't
          touch, but basically a dragonfly's eyes meet and they touch. Damselfly's
          eyes are separate. Ok, so there are three ways we can tell a damselfly from
          a dragonfly. Damselfly: thin, holds its wings along its back, with separate
          eyes. True dragonfly: much chunkier, sticks its wings out and its eyes meet.

First of all, Julia shows the students two pictures of dragonflies and informs them
that there are two members of the dragonfly group (order Odonata): damselfly and
true dragonfly. 'Why do they have different names?' Julia asks the students to look
at the pictures carefully to search for the differences between the two kinds of
insects whilst she gives them some clues. One student (S1) distinguishes a difference and says 'That one is skinny and longer', and because the answer is not clear
and may lead to misunderstanding, Julia rephrases that student's words into a complete sentence. Subsequently, S2 spots another difference regarding the thinness of
the wings, but she has to correct this, saying it is about the wings but not that. Then,
instead of asking for more suggestions about the wings, she tells them the second
different feature, that being that true dragonflies stretch their wings out and damselflies hold theirs along the body. Next, she seeks for other answers about the differences in the images until S3 says 'The eyes are different', and she tells the group
that it is something about the eyes, asking them to look again. Immediately after that
she says 'Can you see the gap between the eyes?' and 'On this true dragonfly can
you see the eyes touch?' thus providing the answer explicitly herself. Finally, she
summarizes the different features of the damselfly and the true dragonfly, rather
than asking the students to do so. This comparison of the true dragonfly and the
damselfly is superficial in nature, because Julia guided the students by only comparing two pictures so as to identify the differences between two species. That is, she
was focusing on naming, describing and distinguishing the two flies so as to make
the comparison. However, it was she who passed the information of how to distinguish the true dragonfly from the damselfly to the students, and perhaps it would
have been better if she had held out and waited for them to volunteer more possible
answers.

Metaphor and analogy are important in scientific as well as everyday common
sense reasoning, and both are powerful resources for the teaching of science. In this
regard, Lemke (1990) argues that science is a human activity which involves 'human
actors and judgments, rivalries and antagonisms, mysteries and surprises, the creative
use of metaphor and analogy' (p. 134). Although both analogies and metaphors
express comparisons and highlights similarities, Duit (1991) distinguishes them in
the following way:

> An analogy explicitly compares the structures of two domains; it indicates identity of parts
> of structures. A metaphor compares implicitly, highlighting features or relational qualities
> that do not coincide in two domains. (p. 651)

That is, metaphor is a figure of speech that uses one thing to mean another and makes a comparison between the two, whereas analogy is a kind of logical argument that provides an insight by comparing an unknown subject to one that is more familiar. However, science education researchers (e.g. Leinhardt, Tittle, & Knutson, 2002; Siegel, Esterly, Callanan, Wright, & Navaro, 2007) tend to employ the terms synonymously in their reports as 'a means for utilizing prior knowledge to gain insight into new observations and to formulate new understandings about a phenomenon' (Zubrowski, 2009, p. 33). The use of analogies and metaphors has been promoted in science education for decades; however, empirical studies have shown that only a few teachers explain ideas or concepts through employing analogy and metaphor (e.g. Dagher, 1995; Treagust, Duit, Joslin, & Lindauer, 1992), and during the field observations, it was noticed that only Mark used them in his explanatory talk.

Episode 1: The Form and Function of a Pitcher Plant

The following example was extracted from Mark's guided visits for the students from the FP School when they are investigating the features of carnivorous plants. He informs the students that the carnivorous plants displayed in the glasshouse are growing in a wet environment, where the soil lacked nutrients, and regarding the issue of how do they get nutrients for surviving in such an environment, he proceeds to pose the rhetorical question 'How do they get minerals?' which is followed by his explanation, in which he uses metaphors and analogies to clarify the forms and functions of the different parts of a North American pitcher plant (*Sarracenia*):

Mark: [points at the pitcher tube] This is basically a leaf which is like a deep dark well full of some liquid. [points at the operculum] This is the little umbrella which keeps the rain out. [points at the peristome] And at the back there's some very sweet stuff. It's quite shiny. It's nectar, a bit like honey.

The analogues 'deep dark well', 'umbrella' and 'honey' are used to explain the structure of *Sarracenia*. That is, for a tiny insect, such as a fly or ladybird, the pitcher tube he explains is like a well, which is deep, dark and full of water at the bottom. The lid or operculum of the *Sarracenia* is analogized to an umbrella, which enables the plant to prevent excess accumulation of rainwater, and the nectar is compared to honey, because the latter is also sticky and sweet.

Episode 2: How a Venus Flytrap Captures Insects

Below, whilst Mark is showing the students a pot of Venus flytraps, he compares their leaves with two pages of a book and subsequently metaphorically models how the plant captures the insect, being as if the pages are shut together.

Mark: This one (Venus flytrap) is a type like two pages of a book. When a fly lands on the surface, it will snap shut and trap the fly inside and it can't get out. [uses hands to simulate the movement of a booking being shut.]

Owning to the limited visiting time that the students had, it was impossible for them to have the direct opportunity to observe the whole predation process of Venus flytraps and the analogue, the closing process of a book, provided the students with a vivid image of how the plant captures insects for minerals. In another workshop for a group Year 3 students, he adopted the same approach to explain the trapping process of Venus flytraps, and when those students were taking their lunch break, before they swapped with the schoolteacher-led group, some children were modelling how the plant trapped insects with their hands.

Episode 3: It's Like a Pint of Beer for Flies
When students ask questions, this usually indicates that they are engaged in the learning process and curious to know more (Chin, Brown, & Bruce, 2002). However, sometimes their questions are difficult to answer by a simple explanation and the use of analogy and metaphor can be helpful. The following is an example how one BGE employed analogy to explain a student's wonderment question.

Mark: [points at the valerian] This is a plant that helps you to sleep. It's called valerian, like a natural sleeping potion.

S9:   Did you remember the sleeping, err there's a plant you showed us, flies eat the sleeping nectar. If you eat it, do you fall asleep?

Mark: Okay. Someone told me that the pitcher plant with a deep dark well eats the flies. We have to drink a pint of that stuff to get sleepy. So it could be considered a bit like beer. So that's the beer for flies. It makes you sleepy after a pint.

After Mark introduces the valerian, a kind of medical plant which can be used for insomnia, he talks about its medical function as a natural sleeping potion. The phrase 'sleeping potion' activates S9's memory of the pitcher plant, which came up in the carnivorous glasshouse before the students arrived at the medical plant bed, where Mark pointed out that insects felt drowsy and fell to the cupped leaf once they sucked the nectar secreted from the plant. S9 connects the medical function of valerian to the pitcher plant nectar and asks whether human beings could fall asleep, if they drank that nectar. In order to give that student a satisfactory answer to such a wonderment question, Mark analogizes the amount of nectar that insects drink to a pint of beer that might make a human being feel sleepy. It is obvious that the pitcher nectar cannot get a human being drunk only by licking the pitcher leaf, but the question generated by that student illustrates that he was highly engaged in learning and thinking. Moreover, through comparing beer and nectar, Mark facilitated the students' understanding of something abstract by pointing out similarities in the real world (Dagher, 1995).

Looking at the episodes presented above, the sources that the BGE used, such as 'deep dark well', 'umbrella', 'two pages of a book' and 'a pint of beer', are familiar messages for young children. However, when educators use analogies and metaphors to teach complex abstract ideas, as Zubrowski (2009) advised, 'students need to become well acquainted with a specific domain if there is a mapping of understandings from this domain to another when there are teacher-introduced analogies' (p. 314). The use of analogies and metaphors in Mark's guided visits provides support what Richard et al. (2006) have proposed, which is that the use of analogies in teaching practice can facilitate students to increase their domain knowledge by shifting from object similarity to relational similarity. However, the analogies and metaphors employed in Mark's guided visits were quite simple compared to those observed in Australian and American secondary science classrooms (Dagher, 1995; Duit, 1991) with older learners, but as Zubrowski (2009) suggested, 'care has to be taken particularly with younger students of the kind and manner in which analogies are introduced and used' (p. 320).

6.3.5.2 Scientific Storytelling

Explanation, as Ogborn et al. (1996) argued, can be thought of as being analogous to telling a story, and hence, in practice, storytelling is an effective approach for helping young children make sense of abstract ideas (Egan, 1986; Wright, 1995). In the current research, storytelling was not commonly identified during the BGEs' teaching practice, but occasionally it arose in their explanatory talk, as shown in the following episodes.

Episode 1: Insects and the Pitcher Plant
The example below relates to a workshop session about carnivorous plants with the students from the FP School, for which the BGE had several things to explain. Predation is a biological interaction where a predator feeds on its prey, and thus, it involves the explanation of how a carnivorous plant lures insects. Furthermore, ideas such as the structures and functions of each part of the plant and what happens to insects are interdependent, in that each needs to be linked to the others in order for the whole predation process to make sense. In the transcript below, the strategy Mark chooses to use is a story that contains a clear explanation of this complex system, thereby simplifying the acquisition of the relevant knowledge.

Mark: What happens is the fly eats some of this sleeping potion. They walk back this way [points at the pitcher mouth] and are about to fly off and what happens is they feel incredibly sleepy and they lose their footing and falling inside the leaf. At the bottom of leaf, like a deep dark well, there is the liquid. They fall into the liquid, [moves his shoulders forwards and backwards] they struggle and they struggle more and they can't get out. There is a fly just coming here, must wish him good luck. There also at the bottom of the leaf are a lot of tiny hairs pointing down. To us they will be just like very fine hairs or something like dog fur, but to an insect it's more like the bottom of a well completely stuck, smooth sided and has a lot of spears pointing down. It's a pretty horrible way to die. They fall into the liquid, can't get out, drown and turn into something most delicious—fly soup. And that fly soup is sucked down by the plant and it helps the plant to grow better.

What Mark is doing is narrating the trapping process in a personalized tone, which is highly memorable, and hence, explaining through storytelling in this way has great potential for visualizing non-observable processes or phenomena. Moreover, the story of how the pitcher plant traps insects may have enhanced the students' engagement with the visit as a whole.

Episode 2: Belladonna and Italian Women
BGEs have the responsibility to introduce medical plants to the visiting schoolchildren, as learning about their uses is an important domain in botany. The excerpt below is extracted from a workshop where the FP School students are on a guided tour of the medical plant bed with Mark.

Mark: If you had surgery on your eyes, the surgeons will look at the back of your eye. They use this plant to make your eyes really widely open so they can see the back much more clearly. Centuries ago Italian women used to put the berry juice of this plant into their eyes to make their eyes open wider and make them look even more beautiful. That is why this plant is called Belladonna, which in Italian means beautiful woman. But the thing is they didn't realize they were poisoning themselves very slowly. Because if you have to take eye drops, if your eyes take eye drops once and then you'll realize that you can taste them about a minute later because our eyes, our ears and our mouth are all connected to the back to our throat. So they were slowly poisoning themselves for the price of beauty.

The story of Italian women who used deadly nightshade berries as cosmetics and eventually poisoned themselves conveyed the fact that this medical plant is a double-edged sword. That is, on the one hand, it can be used in eye surgery as the plant juice can make eyes open widely, whereas on the other, it is poisonous in large doses or when in continuous use. It was observed that the students were highly engaged, keeping very quiet, whilst Mark was telling the story and some even opened their eyes wide or used their fingers to model the image of widely opened eyes.

Episode 3: The Male and Female Parts of Flowers
Flower structure is an important topic in the KS2 science module,[2] as addressed on the Department for Children, Schools and Families Standards Site. The following excerpt is taken from Julia's explanation of flower reproduction to the UW School students.

Julia: So the first person who told me the carpel is made of the stigma, style and ovary. That's the female part of the flower, the reproductive part of the flower. How about the stamen?
S6:   They produce the pollen.
Julia: They do. That's the male part of the flower. So flowers also have male and female parts. When scientists earlier on, in sort of the 18th century, people like Carl Linnaeus were discovering about how plants have male and female parts people were slightly shocked, because they felt the flower is a sort of suitable study for young ladies because there's nothing outrageous in them… there's no… sex doesn't come into the reproduction of flowers.. But sex does come into the reproduction of flowers, because we have to have male and female parts that are kind of very important in terms of making variations in flowers.

---

[2] Unit 5B, life cycles; Section 5, flower parts for reproduction.

Julia informs the students that the carpel and stamen are the female and male parts of the flower. The story is brief, but it presents the idea that the people in the eighteenth century did not understand that the flower's reproduction comes from sex, until Carl Linnaeus, the Swedish botanist, discovered that they have male and female parts. The anecdote narrated by Julia could act as an aide-memoire for the students in relation to the sexuality of plants, and moreover, it also expanded their knowledge about the history of botany.

6.3.6   Summary

The teaching narratives of the BGEs presented above played an important role in enhancing student engagement, facilitating knowledge building and developing conceptual and epistemological understanding of ecology. The detailed pedagogical functions of each observed teaching narrative strategy are set out in Table 6.3.

The teaching narratives observed were predominantly in the form of lecturing, during which the BGEs played an authoritative role that involved their manipulating

Table 6.3   Categorizations of the observed BGE teaching narratives

Teaching narratives	Pedagogical functions
Developing the conceptual line	Talk with the purpose of assisting students to construct knowledge
Shaping ideas	
Uptake ideas	
Shaping the epistemological line	Talk that addresses the epistemological features of botany
Promoting shared meaning	Talk that creates a shared experience which supports students' meaning making during the visit
Presenting ideas to the class	
Sharing the experience of individual students with the class	
Sharing the group findings with the class	
Repeating a student idea to the class	
Using the 'we' statement	
Checking understanding	Talk that seeks students' understanding of the topics addressed during the visit
Checking students previous knowledge	
Checking consensus	
Explanation by comparing	Superficial comparison: talk that focuses on naming, identifying and observing for comparing
Superficial comparison	
Comparing through analogy and metaphor	Comparing by using analogy and metaphor: talk that involves using analogies and metaphors to simplify difficult topics and to make the unfamiliar familiar
Scientific storytelling	A narrative tone of describing and explaining an idea or concept

the interaction with the students. In this regard, in the subsequent interviews arranged to reflect on their teaching practices, all the BGEs stated that it was the time constraints that limited their opportunities for engaging in conversations with the students. Apart the reflections on the issues regarding dialogic teaching, some BGEs also acknowledged that their teaching would be improved if they could embed their explanations in stories, so as to 'bridge the gap between children and the content of science' (Eldredge, 2009, p. 84).

6.4   Mapping the BGEs' Pedagogical Behaviours

As explained above, the observed BGEs' pedagogical behaviours can be grouped into three categories: class management, pedagogical moves and teaching narratives, according to their forms and pedagogical functions. A framework that presents the pedagogical relations among these three categories and interprets how they contribute to students' learning is presented below in Fig. 6.2.

The BGEs' behaviours categorized as class management in the figure above have the aim of providing the students with optimal opportunities to participate in the learning activities. With respect to this, effective group organization can increase the students' opportunity for accessing the facilities offered by the botanic garden setting. Moreover, the BGEs' instruction on health and safety issues and expectations regarding student behaviour can ensure that there is a well-ordered environment for learning. Further, effective collaboration with the schoolteachers can increase the level of the students' participation.

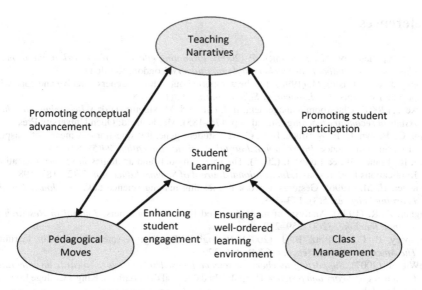

Fig. 6.2 A framework of the BGEs' pedagogical behaviours

Figure 6.2 also shows that the pedagogical moves of the BGEs can enhance the levels of student engagement with the learning activities. For example, in this research it was observed that pointing at particular plants ensured that the children knew which plant was under discussion, whilst demonstrations assisted them in understanding how to manipulate equipment. In addition, the pedagogical moves and teaching narratives were often intermixed so as to help facilitate further the students' conceptual understanding.

The BGEs' teaching narratives, as it is shown in the figure, have the aim of advancing the students' ability to make sense of abstract scientific concepts and processes. The approaches that the BGEs can use to communicate the ecological knowledge to the students, including developing the conceptual line, shaping the epistemological line, promoting shared meaning and checking students' understanding, engage them in developing ideas and when applied effectively will result in high levels of student participation in the lesson discourse. Moreover, explanatory talk, especially storytelling, can transform the delivery of dry content knowledge into a learning experience that all students wish to join in with.

To sum up, in this chapter, the observed BGEs' pedagogical behaviours have been considered in terms of three categories that emerged from the raw data analysis: class management, pedagogical moves and teaching narratives. From this, a framework that describes the optimal pedagogical behaviour for a BGE was put forward. To summarize this, through good class management, they can create a well-ordered learning environment for the students. Moreover, the more opportunities they seize to devise and integrate pedagogical moves with their teaching narratives, the more effective their practice will be.

References

Baines, E., Blatchford, P., & Kutnick, P. (2009). *Promoting effective group work in the primary classroom: A handbook for teachers and practitioners*. London: Routledge.

Brophy, J., & McCaslin, M. (1992). Teachers' report about of how teachers perceive and cope with problem students. *The Elementary School Journal, 93*(1), 2–68.

Burke, J. (2007). Classroom management. In J. Dillon & M. Maguire (Eds.), *Becoming a teacher: Issues in secondary teaching* (3rd ed., pp. 175–185). Maidenhead: Open University Press.

Chin, C., Brown, D. E., & Bruce, B. C. (2002). Student-generated questions: A meaningful aspect of learning in science. *International Journal of Science Education, 24*(5), 521–549.

Coll, R., France, B., & Taylor, I. (2005). The role of models/and analogies in science education: Implications from research. *International Journal of Science Education, 27*(2), 183–198.

Crowder, E. M. (1996). Gestures at work in meaning-making science talk. *The Journal of the Learning Sciences, 5*(3), 173–208.

Dagher, Z. R. (1995). Analysis of analogies used by science teachers. *Journal of Research in Science Teaching, 32*(3), 259–270.

Dempsey, B. C., & Betz, B. J. (2001). Biological drawing: A scientific tool for learning. *The American Biology Teacher, 63*(4), 271–279.

DeWitt, J. (2007). *Supporting teachers on science-focused school trips: Towards an integrated framework of theory and practice* (Unpublished doctoral dissertation). King's College London, London, UK.

Duit, R. (1991). On the role of analogies and metaphors in learning science. *Science Education, 75*(6), 649–672.

Edwards, D., & Mercer, N. (1987). *Common knowledge: The development of understanding in the classroom.* London: Routledge.

Egan, K. (1986). *Teaching as storytelling: An alternative approach to teaching and curriculum in the elementary school.* London: The University of Western Ontario.

Eldredge, N. (2009). To teach science, tell stories. *Issues in Science and Technology, 25*(4), 81–84.

Emmer, E. T., & Stough, L. M. (2001). Classroom management: A critical part of educational psychology, with implications for teacher education. *Educational Psychologist, 36*(2), 103–112.

Evertson, C. M., & Emmer, E. T. (1982). Effective management at the beginning of the school year in junior high classes. *Journal of Educational Psychology, 74*(4), 485–498.

Evertson, C. M., & Weinstein, C. S. (2006). Classroom management as a field of inquiry. In C. M. Evertson & C. S. Weinstein (Eds.), *Handbook of classroom management: Research, practice, and contemporary issues (Vol. 3–16).* Mahwah: Lawrence Erlbaum.

Falk, J. H., & Adelman, L. M. (2003). Investigating the impact of prior knowledge and interest pm aquarium visitor learning. *Journal of Research in Science Teaching, 40*(2), 163–176.

Gilbert, J., Boulter, C. J., & Elmer, R. (2000). Positioning models in science education and in design and technology education. In J. Gilbert & C. J. Boulter (Eds.), *Developing models in science education* (pp. 3–18). Dordrecht: Kluwer.

Griffin, J. M. (1998). *School-museum integrated learning experiences in science: A learning journey* (Unpublished doctoral dissertation). University of Technology, Sydney, Australia.

Hargreaves, L. J. (2005). *Attributes of meaningful field trip experiences* (Unpublished Master's thesis). Simon Fraser University, Vancouver, Canada.

Harlen, W. (2000). *The teaching of science in primary schools* (3rd ed.). London: David Fulton Publishers.

Harlow, A., & Jones, A. (2004). Why students answer TIMSS science test items the way they do. *Research in Science Education, 34*(2), 221–238.

Harrison, A., & Treagust, D. F. (1996). Secondary students' mental models of atoms and molecules: Implications for teaching chemistry. *Science Education, 80*(5), 509–534.

King, H. (2009). *Supporting natural history enquiry in an informal setting: A study of museum explainer practice* (Unpublished doctoral dissertation). King's College London, London, UK.

Kohn, D. (2008). Darwin the botanist. *Roots, 5*(2), 5–8.

Kutnick, P., & Kington, A. (2005). Children's friendships and learning in classrooms: Social interaction and cognitive development? *British Journal of Educational Psychology, 75*(4), 521–538.

Lave, J., & Wenger, E. (1991). *Situated learning: Legitimate peripheral participation.* Cambridge, UK: Cambridge University Press.

Leinhardt, G., Tittle, C., & Knutson, K. (2002). Talking to oneself: Diaries of museum visits. In G. Leinhardt, K. Crowley, & K. Knutson (Eds.), *Learning conversations in museums* (pp. 103–132). Mahwah: Lawrence Erlbaum Associates.

Lemke, J. L. (1990). *Talking science: Language, learning and values.* Norwood: Ablex Publishing.

McNeill, D. (1992). *Hand and mind: What gestures reveal about thought.* Chicago: University of Chicago Press.

Mercer, N. (1995). *The guided construction of knowledge: Talk amongst teachers and learners.* Clevedon: Multilingual Matters.

Nystrand, M. (1997). *Opening dialogue: Understanding the dynamics of language and learning in the English classroom.* New York: Teachers College Press.

O'Connor, M. C., & Michaels, S. (1996). Shifting participant frameworks: Orchestrating thinking practices in group discussion. In D. Hicks (Ed.), *Discourse, learning, and schools* (pp. 63–103). Cambridge, UK: Cambridge University Press.

Ogborn, J., Kress, G., Martins, I., & McGillicuddy, K. (1996). *Explaining science in the classroom.* Buckingham: Open University Press.

Osborne, J., & Hennessy, S. (2003). *Science education and the role of ICT: Promise, problems, and future directions*. Bristol: Futurelab.

Richard, L. E., Morrison, R. G., & Holyoak, K. J. (2006). Children's development of analogical reasoning: Insights from scene analogy problems. *Journal of Experimental Child Psychology, 94*(3), 249–273.

Roth, W.-M., & Lawless, D. (2002). Scientific investigations, metaphorical gestures, and the emergence of abstract scientific concepts. *Learning and Instruction, 12*(3), 285–304.

Scott, P. H. (1998). Teacher talk and meaning making in science classrooms: A Vygotskian analysis and review. *Studies in Science Education, 32*(1), 45–80.

Siegel, D. R., Esterly, J., Callanan, M. A., Wright, R., & Navaro, R. (2007). Conversations about science across activities in Mexican-descent families. *International Journal of Science Education, 29*(12), 1447–1466.

Stavrova, O., & Urhahne, D. (2010). Modification of a school programme in the Deutsches museum to enhance students' attitudes and understanding. *International Journal of Science Education, 32*(17), 2291–2310.

Treagust, D. F., Duit, R., Joslin, P., & Lindauer, I. (1992). Science teachers' use of analogies: Observations from classroom practice. *International Journal of Science Education, 14*(4), 413–422.

Vergou, A. (2010). *An exploration of botanic garden-school collaborations and student learning experiences* (Unpublished doctoral dissertation). University of Bath, Bath, UK.

Vygotsky, L. S. (1987). Thinking and speech (N. Minick, Trans.). In R. W. Rieber, A. S. Carton, & J. S. Bruner (Eds.), *The collected works of L.S. Vygotsky: Problems of general psychology* (Vol. 1). London: Plenum.

Wells, G. (1999). *Dialogic inquiry: Towards a sociocultural practice and theory of education*. Cambridge, UK: Cambridge University Press.

Wertsch, J. V. (1991). *Voices of the mind: A sociocultural approach to mediated action*. London: Harvester Wheatsheaf.

Wright, A. (1995). *Storytelling with children*. Oxford: Oxford University Press.

Zubrowski, B. (2009). *Exploration and meaning making in the learning of science*. Dordrecht: Springer.

Chapter 7
Discussion, Reflections and Implications

The aim of this study has been to investigate the pedagogical practices of BGEs through a consideration of their guiding school visits aimed at engaging and supporting children's learning of ecological science. As explained in Chap. 1, the rationale for conducting this research is that there is a gap in the literature regarding the nature of informal educator-guided school visits to botanic gardens. The practical outcomes of this study have allowed for an in-depth examination of the structure of guided visits and BGEs' teaching practices, which addresses the aforementioned gap in the literature regarding school trips to botanic gardens and pedagogy in these informal contexts. Moreover, this contribution is important because it offers a close examination of the teaching and learning processes in an out-of-school setting that can assist schoolteachers' preparations for conducting botanic garden visits. Further, the identification of the BGEs' pedagogical behaviours has led to the development of a framework, which can aid or assist researchers in describing informal science educators' practices as well as in carrying out further investigation in the field. Moreover, through carrying out this extensive in-depth fieldwork, it has been demonstrated that combined data collection methods can not only be employed in school classrooms (Stigler, Gonzales, Kawanka, Knoll, & Serrano, 1999) and/or museum sites (Allen, 2002; DeWitt & Osborne, 2007) but can also contribute to gathering rich data in complicated outdoor settings such as botanic gardens. The outcomes of this study enrich the understanding of the sociocultural theory of teaching and learning with respect to informal contexts, particularly by highlighting the importance of the interactions between the BGEs and students. More specifically, the study of the discourse that occurred during the guided visits enabled me to examine the detailed processes, wherein the BGEs supported and facilitated the children's meaning making of ecological science. In this final chapter, the key findings are discussed in order to address the research questions (Sect. 7.1), and the practical and the theoretical implications of this study are also presented (Sect. 7.2). The limitations of the research are described and some directions for future research undertakings are identified (Sect. 7.3).

© Springer Science+Business Media Singapore 2015
J. Zhai, *Teaching Science in Out-of-School Settings*,
DOI 10.1007/978-981-287-591-4_7

7.1   Discussion of the Key Findings

7.1.1   The Structure of the Guided Visit

The data gathered in this study have shown that the BGEs' guided school visits were learning oriented, although formal lecturing took up a large part of their teaching time. From the in-depth investigation of the BGEs' pedagogical behaviours in Chap. 6, it has emerged that the lecturing type of talk contained storytelling, analogies and metaphors and models, which made the explanations more interesting to the young children. Similar findings have not been widely reported in other studies regarding school trips in informal settings (Gilbert & Priest, 1997), although some museum education researchers have emphasized the role of storytelling and analogies in generating personal connections with exhibits (Anderson, Piscitelli, Weier, Everett, & Tayler, 2002; Bedford, 2001). Moreover, Cox-Peterson, Marsh, Kisiel and Melber (2003) have investigated museum educators' guided school tours in natural history museums and reported that more than 60 % of the observed visits involved the use of scientific jargon. In contrast, in this current research it was found that the BGEs sometimes used analogies or models to help them interpret the meaning of complex science concepts for the students rather than simply telling them the scientific terms.

In a relatively recent study regarding school visits to museums in Israel, Tal and Morag (2007) found that the activities entailed in museum educator-guided tours were generally passive directed and involved very short periods of student voluntary exploration. However, in contrast with this, the school trips to botanic gardens observed in this study were largely taken up with guided-exploration and active-directed activities, with the one exception being Debbie's lessons, which centred on PowerPoint-based presentations. The introductory talk given by the BGEs was relatively short compared to that reported by Tal and Morag (2007) in their study, but even so, a number of students still complained that the orientation session was the least interesting part of the visit. Nevertheless, the briefing session about the health and safety issues is unavoidable because it contains important information. That is, although the students usually may well feel that listening to the BGE's introductory talk is boring, a thorough review of outdoor learning in England and elsewhere has shown that health and safety issues are the major concern for schools when organizing trips (Rickinson, Dillon, Teamey, Morris, & Choi. et al., 2004).

Most of the visits that were observed for this study combined indoor and outdoor sessions, which is consistent with the natural setting of botanic garden education. The BGEs usually started the lesson inside the classroom, where they checked the students' previous knowledge and introduced the schedule for the guided visit. During this session the students could review and develop their basic understanding of ecological science in preparation for the learning experiences later on during the visit (Griffin & Symington, 1997). For example, during Simon's guided visits, he reviewed students' knowledge of plant structures and the notion that plants need to grow, before the group went out to explore the glasshouses and find answers to the questions on their worksheets. Similarly, the other BGEs would prepare the students inside the classroom prior to their garden exploration.

When compared with students following museum educator- or schoolteacher-guided school tours (Cox-Petersen, Marsh, Kisiel, & Melber, 2003; Griffin, 2004; Griffin & Symington, 1997; Tal & Morag, 2007), those in this study had more opportunities to explore and investigate the sites, either by themselves or with the guidance of the BGEs. More specifically, activities, such as making observational drawings of plants, identifying pond life and searching for minibeasts, encouraged the students to extend their scientific enquiry skills, whilst at the same time fostering their curiosity about the natural world. These points were reflected in students' completed impression sheets, where they stated that the active-directed activities were the most memorable and that they were interested in understanding more about the lives of organisms in different ecosystems.

The participating BGEs in this study were experienced outdoor practitioners, clearly familiar with their working environments. Consequently, they appeared to be more confident in directing students and organizing the various outdoor learning activities than their school-based counterparts, which is consistent with some previous research that has elicited that schoolteachers know little about teaching outside the classroom and this often results in their failure to facilitate children's learning in an effective and efficient way in such settings. The data collected for this study has indicated that the guided visits were designed by the BGEs according to the availability of the facilities and resources in the botanic gardens, which expresses the aim of optimally connecting the content of the school trips to the National Curriculum. Even though the BGEs had to channel students from one learning station to another during a visit, the moving only took up a small proportion of the entire visiting time, which allowed the students plenty of time to engage themselves with the hands-on exploratory activities. Nevertheless, despite this careful management of the visits by the BGEs, some students still complained that they talked too much and reduced the time they had to explore the garden by themselves.

Apart from giving the students sufficient opportunities to be actively directed in the activities by themselves, the BGEs supported the children's explorations by giving additional information, giving demonstrations and providing supervision. During group work or individual exploration, it was difficult for one individual BGE to help all the students in the class. Therefore, the BGEs sometimes encouraged the schoolteachers to participate in students' activities to assist them with learning. Moreover, a large proportion of the guided-exploration sessions in the observed lessons involved a child-centred learning environment, thus increasing the opportunities for the students to interact with adults, which contributed to the social construction of shared meanings (Edwards & Mercer, 1987).

A combination of active-, passive- and guided-directed activities during the observed visits was the main structure of the BGE-led school trips, with the exception being that of Debbie's lessons that were mainly passive directed and left limited space for the students to direct their own learning. Bamberger and Tal (2007) have found that 'limited choice' visits, that is, those in which students are allowed to control their learning as well as benefit from the scaffolding offered from the educator, are more engaging than highly structured or unstructured visits. Moreover, DeWitt and Storksdieck (2008) propose that a structured experience can enhance cognitive learning, but may dampen interest or result in less positive attitudes among learners. Similarly, in a German study of fifth graders' learning experience on a visit to a natural history museum, Wilde and Urhahne (2008) found that children's intrinsic motivation was relatively higher with open tasks, although a certain amount of instruction could be appropriate for primary aged children.

7.1.2   The BGEs' Views on Pedagogies in Informal Settings

In their interviews, when asked specifically about the pedagogical perspective informing their practice, all the BGEs referred to learning through hands-on activities, building knowledge upon prior experiences and constructing shared understanding during social interactions. It would appear, therefore, that these BGEs' views on pedagogy can be characterized as falling within the constructivist perspective, which holds that:

> Learners are thought of as building mental representations of the world around them that are used to interpret new situations and to guide actions in them. These mental representations or conceptual schemes in turn are reviewed in the light of their fit with experience. Learning is thus seen as an adaptive process, one in which the learners' conceptual schemes are progressively reconstructed so that they are in keeping with a continually widening range of experience and ideas. It is also seen as a creative process of 'sense making' over which the learner has some control. (Driver, 1989, pp. 481–482)

The proponents of constructivism acknowledge the need for an active, central role for learners in the education process. Moreover, under this perspective the educator is not only constantly required to assess the knowledge that the children have gained but also to facilitate their learning through modelling, coaching and scaffolding (Jonassen, 2004).

Learning through hands-on experience has been widely advocated under the constructivist lens for a variety of different contexts (e.g. Hein, 1998; Tenenbaum, Naidu, Jegede, & Austin, 2001; Windschitl, 2002). With respect to this, in their interviews the BGEs highlighted the role of hands-on activities in enriching the children's visit experiences. For example, Mark believed that encouraging the children to use their sensory modalities to interact with authentic specimens could provoke curiosity and, thus, enhance their motivation for learning. Similarly, Julia allocated a particular role to the children, namely, that of being scientists, which required them to explore the biological world through first-hand experiences, such as making observational drawings, pond dipping and minibeast hunting, whilst Simon required the students to observe and explore things so as to be able to answer the questions listed on their worksheets.

Many students reported that the sessions that gave them opportunities to directly interact with nature, by using their multisensory modalities, were the most impressive. In this regard, the collected data have shown that the students were actively engaged during the hands-on activities in that they were discussing with their peers, posing questions to the BGE and so forth. This finding is in line with that of Paris, Yambor and Packard (1998), who reported that hands-on tasks during biology sessions have a positive influence on primary students' level of interest in science and their abilities with regard to problem-solving. Moreover, Kellert (2002) argues that experience in the outdoors has the potential to promote children's 'interest, curiosity and capacity for assimilating knowledge and understanding of the natural world' (p. 133), whilst Carson (1998) claims that such experiences can serve as powerful motivations and stimuli for learning and development. However, Osborne (1998) argues that 'experience, of itself, while highly enjoyable, is overwhelmingly a missed learning opportunity without some attempt to encourage the visitor to focus, recapitulate and review' (p. 9). That is, the outcome for children who participate in an educational excursion to informal settings should be geared towards achieving cognitive and affective gains (DeWitt & Storksdieck, 2008). In order to promote conceptual understanding and emotional development, some museum researchers have advised that effective school trips should include a moderate amount of structure, whilst still allowing for free exploration (Falk & Dierking, 1992; Price & Hein, 1991; Stavrova & Urhahne, 2010). This study proposes that the hands-on experience should not be the goal of the botanic garden visit, but should serve as a vessel or medium through which to facilitate children's sense making of the natural world.

The role of social interaction in engaging and supporting children's learning was acknowledged by the BGEs during the interviews. More specifically, they agreed that knowledge is constructed by students when they are communicating and sharing ideas with their peers and adults. Both Mark and Julia went on to suggest that encouraging students to work in small groups and giving them opportunities to talk to each other were important, and in his responses Simon emphasized the role of questioning for enhancing the quality of educator-learner interaction. Debbie expressed the view that she liked to get the accompanying schoolteachers involved in class discussions, because this helped to show them that it was an important learning opportunity and, hence, the children were more likely to participate. Although most of the BGEs expressed in their interviews that they would take the students' different ages and backgrounds into account when organizing the lessons, generally, they failed to modify their approach according to the different group attributes. Moreover, it was observed that experienced difficulties balancing the needs of the group as a whole with those of individuals. Furthermore, not all the BGEs communicated the social, historical and cultural perspectives of plants, but instead tended to focus more on presenting scientific concepts.

7.1.3   BGEs' Pedagogical Practices in Informal Settings

This study set out to investigate the pedagogical moves of BGEs from a discourse perspective. The findings of this study suggest that, although their discourse predominated during the guided visits, the diverse communicative approaches adopted and the variety of questions asked revealed the complexity of their pedagogical practices. Firstly, the analysis based on the four classes of communicative approach has revealed that the BGE discourse was interactive and authoritative in nature. Despite the fact that interactive/dialogic discourse was occasionally observed, it was overwhelmed by either an interactive/authoritative or noninteractive/authoritative pattern. Mortimer and Scott (2003) point out that in dialogic discourse 'more than one point of the view is represented, and ideas are explored and developed, rather than it being produced by a group of people or by a solitary individual' (p. 34). In this regard, the students who participated in the BGEs' guided visits were not given adequate opportunities to represent and explain their different points of view. Instead, their participation in lesson discourse was restricted and controlled by the BGEs. The frequent use of 'evaluate' and 'insert' moves when the BGEs responded to the contributions made by their students was the supporting evidence for this view. These findings are in line with what previous research on the features of classroom talk has found in that teachers generally maintain authority and control the direction of talk by making evaluative comments (Cazden, 2001) or offering direct explanations as they cannot avoid 'telling' (Lobato, Clarke, & Ellis, 2005). Moreover, it would appear that these pedagogical moves act to remind students who is ultimately in control of the lesson discourse.

Despite the fact that the BGEs manipulated the discourse of the observed guided visits, it was noted that they did sometimes mediate students' making sense of the science content through dialogic interactions. For example, they were found using 'maintain', 'elicit', 'press', 'revoice' and 'repeat' moves when responding to their students' talk. There is abundant evidence that such follow-up moves have the potential for encouraging student participation by sharing their views and thus promoting dialogic discourse (Edwards & Mercer, 1987; Nystrand, 1997). Furthermore, when the BGEs used open-ended questions that invited responses, the lesson discourse opened up beyond the traditional lecture format of teaching by telling. These findings contrast with some previous research investigating school trips to museums during which informal educators asked a large number of questions without follow-up, elaboration or probing (Cox-Petersen et al., 2003; Tal & Morag, 2007).

The analysis of the BGEs' lecture format of discourse found that scientific ideas were often addressed through storytelling and the use of analogies. For instance, Mark's story of the deadly nightshade berry juice as a cosmetic for Italian women would have assisted students in learning about how the use of medicinal plants could be dangerous. Similarly, Julia's story of Carl Linnaeus might have reinforced the students' memory about the reproduction of flowers and enriched their knowledge about the history and development of botany. Matthews (1989) has advocated a historical and philosophical approach to the teaching of school science which, it is claimed, would contribute to a better and more effective conveyance of the nature of the subject. The historical components embedded within Mark and Julia's storytelling would have allowed the students to imagine another time and place 'to create their own meaning and find the place, the intersection between the familiar and the unknown' (Bedford, 2001, p. 33). In general, the findings support the argument that storytelling is an important means for science communication to convey information in an accurate, attractive, imaginative and memorable way (Klassen, 2009; Negrete & Lartigue, 2004).

In addition to storytelling, the use of analogies was observed during Mark's lecturing format of discourse through which he made the explanation of carnivorous plants interesting and the unobservable biological process visible. The observations from the present study would suggest that the use of analogies should be considered as meaningful practice to engage students by generating personal connections with the content. For instance, Mark used 'umbrella', 'deep dark well' and 'honey' to describe the features of a pitcher plant, which may have facilitated the students as his approach related the content to their daily life knowledge, and thus, they could easily make sense of the new information. However, despite Mark and Julia using analogy and storytelling, they both maintained an authoritative role throughout their talk without resorting to any dialogic interactions with the students. Under these circumstances, 'the most fluent exponent of scientific ideas does all of the talking whilst the novices have little or no opportunity to speak the scientific language for themselves and to make it their own' (Scott, Mortimer, & Aguiar, 2006, p. 622). That is, if we want students to engage in meaningful learning, they need to be provided the opportunities to make sense of newly learned knowledge through their own talk, but on the whole, from the observations, it was apparent that this form of interactive pedagogical practice was not being employed. The sight of students gesturing when Mark used analogies to explain a biological process gives credence to the view that he could seize on this to involve the learners in a dialogue that would embed the knowledge acquired more deeply. In sum, I would argue strongly that the BGEs need to balance their authoritative talk with dialogue, as 'both dialogic and authoritative discourse have critical and complementary functions in supporting student learning' (DeWitt & Hohenstein, 2010, p. 456).

Furthermore, I noted that some of the lesson discourse was driven by the students' volunteering talk, indicating that they can reverse the interactional roles, change the situational power asymmetry and wield power to control the discursive interaction (DeWitt & Hohenstein, 2010). The analysis of this student volunteering talk suggests that the goal of the activity, task oriented or knowledge oriented, might have an impact on what they would like to address when they initiate the interaction with the BGEs. Moreover, apart from the sociocultural differences, such as language barriers, familiarity with informal settings and confidence in speaking in public, whether students can take the initiative in the lesson discourse would also appear to relate to the communicative approach of the BGEs. For example, the students taught by Simon generated the least amount of volunteering talk among the three BGEs. This finding can probably be attributed to the fact that unlike the other groups, his students were visiting a botanic garden for the first time. Moreover, the majority of the students from the SB School were Pakistani and Bengali immigrants who spoke English as a second language. Therefore, the unfamiliarity with the use of the language could have been a barrier for them playing an active role in the discourse (Cuevas, Lee, Hart, & Deaktor, 2005; Wellington & Osborne, 2001), by, for instance, initiating discussion or posing questions.

There is no doubt that teaching science in informal settings, such as botanic gardens, is a challenging task. It requires educators to be familiar with students' school learning experience, sociocultural background, the needs of the individual learner and so forth (Cox-Petersen et al., 2003; DeWitt & Storksdieck, 2008; Tal & Morag, 2007). In sociocultural learning environments, as Ash and Wells (2006) have suggested, in order to move towards greater individual understanding, students should be encouraged to participate in knowledge building by sharing what they know and by providing arguments through the ongoing challenge of responding to other speakers. However, whether this outcome can be achieved depends on the content being taught and the type of mediation provided by educators (Tal, 2012). Thus, I believe that the BGEs might benefit from making the content less informative and more exploratory when designing learning activities. They need to bear the principles of learning suggested by the sociocultural theory in mind and engage students in joint negotiations through dialoguing. In particular, they might strive to create real discussions, in which the educator and students are in equal or similar positions, by shifting the power relation in favour of the latter.

7.2 Implications of the Study

This study of BGEs and schoolchildren at botanic gardens has implications for several groups of stakeholders involved in informal science learning environments and with improving students' experiences of these, including the BGEs themselves, teachers, policymakers and educational researchers. In the following sessions the issues of interest that can be taken from this study by each of these stakeholders are outlined.

7.2.1 Implications for Practices

7.2.1.1 The Design of Education Programme

As noted in the impression sheets completed by the visiting schoolchildren, a large number of students had never been to a botanic garden before. For these students, their level of their sense of novelty relating to the botanical environment was comparatively higher than those who had been before. In their interviews, the BGEs expressed the view that they preferred to tailor their education programmes according to the nature of the different visiting school groups. In practice, however, they found it was difficult to provide every group with a customized programme owing staffing limitations and the high density of visiting school groups in each academic year, especially during the spring and summer terms. Moreover, schoolteachers' purposes for organizing trips to informal institutes vary, in that some have very strong educational objectives, whereas others may consider it as a fun day out for their students (Kisiel, 2005). With regard to this, it is posited that when designing education programmes, the BGEs should classify the visits they have to accommodate into two categories: the general visit and the focused visit (see Table 7.1).

The general visit, as set out above, is designed for those students who have not had much experience of botanic gardens or who attend a visit without having any particular educational objective set beforehand. According to the findings from this study, guided exploration is recommended as the main activity for this group of

Table 7.1 A proposed framework for informing the design of an effective learning experience (for primary school students) in botanic garden settings

	General visit (one-off visit)	Focused visit (a series of visits)
Target group	First time/infrequent visitors	Frequent visitors
	Visitors without specific educational objectivities	Visitors with particular educational objectivities
Objectives	Obtaining a brief conception about the botanical environment	Linking and extending subject matter in school curricula
	Linking daily life experiences with ecological science	Promoting scientific thinking (explanations, argument and decisions)
		Developing basic practical and enquiry skills
Activities	Guided tour: to see different parts of the botanic garden	Collaborating with group members to plan an approach to solve the questions given by the BGE
	Guided exploration: hands-on experiments, observations with appropriate guidance from the BGE	Selecting and managing the data for the questions
	Game, role play: to awaken curiosity and enthusiasm	Developing explanations and arguments to present findings
Teaching strategies	Using open-ended questions to challenge students' thinking	Encouraging students to collaborate and discuss for decision-making and problem-solving
	Promoting dialogic exchange of ideas with children	Facilitating group discussion, argumentation and debate for scientific reasoning
	Integrating storytelling and metaphorical and analogical modelling into explanatory talk	Engaging students with dialogue and asking questions to prompt their thinking
	Giving children free choices to explore when health and safety conditions permit	Using mini-lectures to develop the conceptual and epistemological line
	Being patient and listening to student comments and questions	
Outcome	The botanic garden experience	Appreciating how science works in the botanic garden
	Flow learning	Engendering cognitive and affective development

children, and in addition, games and role playing,[1] which were passed over by the participating BGEs in this study, can be adopted as powerful approaches with the potential to awaken curiosity and enthusiasm of young children, especially in the outdoor environment (Cornell, 1989; Kuh, Ponte, & Chau, 2010; Malone, 2007). Playing games, as Cornell (1989) has argued, can transpose learning into a fun, immediate and dynamic experience. Perhaps, the BGEs from the botanic gardens could integrate different storylines into the format of the general visit in order to awaken the children's enthusiasm and inspire their learning.

Turning to the focused visit format, this is designed for the frequently visiting school group or the school visit that has predetermined and specific educational goals set by the teachers. This type of visit should emphasize the close connection between the school subject matter and the facilities available in the botanic garden. As a complement and extension to the school curriculum, the focused visit can facilitate the development of practical skills and scientific thinking through problem-based, enquiry learning. With respect to this, the observations from this study indicated that although some of the BGEs encouraged children to work in groups for exploration, they did not set up 'ground rules' (Edwards & Mercer, 1987) for discussions, which hindered their ability to develop the students' scientific reasoning and problem-solving skills effectively. If they took some time at the start of a discussion to explain to the children what was expected of them, then some 'epistemic' tasks, such as describing, arguing, debating, explaining, predicting, critiquing and explicating, as identified by Ohlsson (1996), would be more likely to emerge throughout the visit.

7.2.1.2 Engagement with and Support for Learning

From the analysis of the BGE-student interactions during the observed visits, it became clear that BGE-mediated experiences have the potential to enable student engagement in the process of scientific enquiry. The data in Chap. 6 was presented in the three key categories that emerged from the raw analysis, and therefore, it is posited that it is the most appropriate schema for considering how to enhance learning engagement, these categories being class management, pedagogical moves and teaching narratives.

---

[1] My visit to the Singapore Botanic Gardens in November 2009 strengthened my perceptions of the games and role play for children's environmental learning. The guided tour that I observed in Singapore was designed for 6–12-year-old children focusing on developing the general identification skills of tropical plants. In contrast to the visits observed in England, this tour integrated role play and games into a storyline—Sarah's birthday. All the children were asked to act as Sarah's friends who were looking for sources in the garden to make a 'birthday gift' for Sarah. During the tour, the BGE presented different plants to the children and encouraged them to collect some samples for a piece of art work. Then, the children were given some time to make the 'birthday gift' with the material they had collected during the tour. In the end, the children played a game (guessing the name of plants according to the 'birthday gift' they had made) to celebrate Sarah's birthday. During my observation, I noted that the children were highly engaged and posed a lot of questions to the BGE.

Hands-on experiences that engage children's multisensory modalities facilitate the construction of an individual's perceptual understanding of an object or a phenomenon (Zubrowski, 2009). According to the data collected on the impression sheets, through which the students reported on their visiting experiences, the facts that they obtained during the hands-on activities were considered by them as being the most interesting and memorable. Hence, it is reasonable to suggest that having an opportunity to touch and observe a wide range of rare and awe-inspiring specimens, which cannot be seen during run-of-the-mill school lessons, can stimulate their level of curiosity and develop their interest in learning about science. Moreover, the hands-on activities provide more spaces for socialization and foster peer collaboration, which not only leads to cognitive gains but is also beneficial in terms of more positive student attitudes in the form of increased levels of intrinsic motivation and perceived competence (Paris et al., 1998). This outcome is in keeping with Russell (1990), who claims that making the visit most enjoyable and enhancing the students' positive attitudes towards learning science should be the primary goal of informal institutions.

In promoting conceptual advancement, the use of teaching narratives, as discussed in Sect. 6.3, has been shown to be an effective approach towards getting the children more intellectually engaged. In particular, storytelling, analogies and metaphors are tools that can effectively mediate and support children's understanding of complex and abstract ideas. Moreover, the BGEs' epistemological talk, for instance, that given in the example of the introduction to binominal nomenclature, can be used to explain complex matters. This would suggest that young children are capable of acquiring higher cognitive knowledge when they are afforded appropriate support by scaffolding by the educators (Scott, 1998; Zubrowski, 2009).

The analysis of the discourse data has found that the nature of the BGE-student interaction during whole class instruction took the triadic I-R-F pattern, but utterances in chained triadic pattern were seldom observed. Nevertheless, even though the exchanges between the BGEs and the students were invariably short, there were a sufficient number of incidences to show their engagement in science enquiry. That is a good outcome, because previous research on visitors' talk in informal settings, such as museums, science centres and botanic gardens, has found that when children are engaged over an exhibit with adults, their explorations have been observed to be longer, broader and more focused on relevant comparisons, than for those children who are involved with an exhibit without adults (Crowley, Callanan, Jipson, Galco, & Topping et al., 2001; Tunnicliffe, 2001). In sum, from the current research it has emerged that extra time should be devoted to more guided enquiry so as to develop ongoing dialogue between schoolteacher and students or informal educator and students, which would have to come at the expense of time spent on lecturing. By doing so, this would provide the learners with the opportunity to confirm or amend their ideas, thereby engaging them in a process of deep thinking (Smithenry, 2010).

A further implication resulting from this study relates to the finding that questioning appeared a poorly used strategy by most of the BGEs. More specifically, whilst Mark based his instruction around open-ended questions so as to prompt and elaborate students' enquiries, the other BGEs largely posed questions in order to ascertain the children's prior knowledge. In view of this and in accordance with the sociocultural theories of learning that have highlighted the importance of interactions for meaning making, BGEs and other practitioners (informal educators and schoolteachers) are advised that one way through which they could help students concentrate on enquiring would be to ask questions that are open ended in nature and promote productive thinking (Chin, 2007; Martin, Brown, & Russell, 1991; Nystrand, Wu, Gamoran, Zeiser, & Long, 2003).

Although most of the class interactions were initiated by the BGEs, student-generated questions and comments were evident in the observed visits. According to DeWitt and Hohenstein (2010), student volunteering talk is a feature of discourse during museum visits that distinguishes it from that in school classrooms. Moreover, they have posited that the students enjoy a higher level of autonomy and have more power to control their discourse in informal settings. Although there are many contextual factors, such as the backgrounds of the visiting schoolchildren, the objectives and the content of the visit and the preparation work carried out prior to the visit, all of which influence a student's self-efficacy to initiate and engage in a conversation, the BGEs' attitudes towards student-initiated talk are important. Compared to the visits guided by the other BGEs observed for this research, the proportion of student-initiated interactions in Mark's lessons was much greater. In this regard, when transcribing the class discourse recorded from his guided visits, it was noticed that he gave the children who intended to speak enough time to express their ideas in full and, as long as the students were making comments or asking questions, he responded to them with utterances, such as 'yeah', 'right' and 'hmmm', to show that he was listening. Moreover, it was noted that according to the responses given in the impression sheets, some of the children reported that they liked Mark's teaching specifically because they thought he liked listening to their ideas. From this outcome, it is suggested that educators should be encouraged to give students sufficient time to construct their responses, rather than providing the answers themselves, if they are not already doing so.

To sum up, this research was carried out with the purpose of investigating the pedagogical practices of botanic garden educators, especially the techniques they employed to support and enrich the schoolchildren's visit experiences. As noted in Chap. 6, their identified techniques, involving class management, pedagogical moves and teaching narratives, were effective in engaging the students with learning activities. The class management techniques, such as grouping children, assembling groups and monitoring children's behaviour, are helpful in establishing a well-ordered learning environment so that learning activities can be carried out more efficiently. Teaching narratives have the potential to develop the students' conceptual understanding by engaging them in exchanging ideas through dialogue as well as communicating abstract concepts through storytelling. Although a number of incidences of direct instruction or telling were noted in the analysis of BGE-student interactions, some events, storytelling in particular, were found to be effective in conveying a body of content knowledge during a short visit. The pedagogical moves, especially metaphorical modelling, noted in the BGEs' teaching practice can facilitate the students' making sense of unobservable or abstract science processes. For example, Mark used his hand to model how a sundew plant uses its leaves to capture insects for nutrients when he was explaining its predation process. By so doing, the children could easily form an image in their mind about such a process, which cannot be observed in real time during a relatively short visit.

7.2.1.3 Facilitating Collaboration Between School and Informal Science Institution

In order to organize the visit in an effective way, it has generally been accepted that school trips to informal settings need to be planned as an integral part of the curriculum, rather than as an isolated activity or merely as a form of enrichment (Hofstein & Rosenfeld, 1996). The focused visit, as explained in the above section, requires a solid collaborative partnership between staff at the botanic gardens and schools. In a recently completed doctoral research, Vergou (2010) conducted an ethnographic study of the collaborations between Wakehurst Place Garden in West Sussex and local primary schools. The author noted that the level of garden-school collaboration has an influence on the visiting schoolchildren's learning experience and highlighted the importance of the BGE-schoolteacher partnership in preparing, delivering and following up the visit. With regard to this current study, during the UW School visit, which focused on Darwin, plant adaptations and seeds, the BGE, Julia, worked collaboratively with the schoolteacher to support the children's learning:

> This year we had already had a talk and demonstration about Charles Darwin by Julia when she came into school. The children were really interested to link the gardens to her talk about Darwin. After the visit the children talked in groups about what surprised them and interested them about their visit. There was new information as there were so many different plants and seeds to think about, especially the insectivorous plants and those that man uses for clothing and food. Our expectations were more than met, considering that all the children had one seedling to take home and then Julia brought each one a Sundew plant into school later that week. (The email from the schoolteacher two weeks after the visit, received on 15/4/09)

According to both the extant literature and the findings of this study, it appears that the successful collaboration between informal science institutions and schools gives rise to better learning outcomes for the children. Given this, it has still to be determined how informal science institutions should develop partnerships with schools, and thus, a framework for collaborative partnership between them is put forward.

Phase I involves the pre-visit collaboration between the schoolteachers and informal educators, during which they discuss the content and schedule of the visit. The schoolteachers should explain the purpose of the visit and the current academic progress of their students, so that the informal educators can tailor the education programme according to the needs of the school group. These recommendations for the activities undertaken in Phase I are supported by museum researchers, who have called for appropriate pre-visit preparation that provides the students with the necessary support for engaging in appropriate and positive behaviours during the visit (DeWitt & Osborne, 2007; Falk & Balling, 1982; Griffin, 1998). During the visit, the schoolteachers are encouraged to participate in the student activities and support their learning. Subsequently, Phase III refers to the post-visit collaboration, which is considered as the gateway through which the informal science institutions can eventually develop a solid enduring partnership with the schools. With regard to the third phase, in their interviews the BGEs in this study stated that they believed that it was the schoolteachers' responsibility to follow up after the visit. However, a proposal from the outcomes of this research is that if the informal educators could offer some suggestions and assistance regarding follow-up activities, the schoolteachers might make more use of the visiting experience and more of the collaborations could progress to phases III and IV.

With regard to Phase IV, Ramey-Gassert and Walberg (1994) argue that long-term, sustainable collaborations between schools and informal institutions can best meet the needs of both teachers and students. To achieve this, the informal educators need to modify their education programmes so as to meet the requirements stated in the new National Curriculum, whilst maintaining some elements that go beyond the mere fulfilling of this so as to ensure that part of visit involves activities that could not be experienced in classroom settings. Moreover, there is strong current political support for learning outside the classroom where it is proposed that a wider range of people, including parents, voluntary organizations and carers, should become engaged in providing learning opportunities (DfES, 2006). Phase IV marks an enhanced collaborative partnership where the schoolteacher should systematically follow up students' scientific enquiry and reasoning in the school classroom after a series of well-planned focused visits to informal science institutions, with the aid of informal educators. That is, they should both be proactive in promoting cooperation among such groups by holding meetings, delivering training, creating contact lists, fundraising and/or setting up open days, for example. In sum, Phase IV not only involves enhanced collaborative partnership, where schoolteachers can systematically follow up students' scientific enquiry and reasoning developed in the school classroom by working with the informal educators to create well prepared and up to date visits, but also drawing in a wider range of interested parties who could contribute their own expertise to the children's learning, regarding such issues as environmental awareness and conservation.

7.2.1.4 Professional Roles of a BGE

Drawing on the outcomes from the analysis of each individual BGE's pedagogical practice, in this study a range of key professional roles that qualified BGEs should possess have emerged, these being science educator, garden guide, botanist and environmental educator. First, as a science educator, a BGE needs to have sufficient knowledge about pedagogy so as to be able to deliver effective teaching that enhances students' learning experience. As a garden guide, a BGE should be familiar with the setting and be able to encourage visitors to develop a keener awareness, appreciation and understanding of the value of plants in their everyday lives through effective interpretations. In order to communicate ideas and concepts to the audiences in botanic garden settings, Grenier (2009) suggests that a guide's interpretation should provide a moving and meaningful experience, primarily in the place to be interpreted. As discussed in Chap. 6, some of the BGEs' interpretation strategies, such as the use of storytelling, demonstrations, guided walk and talks, proved to be effective in motivating and inspiring students.

Furthermore, a BGE should be an expert in botany so that she/he can provide professional information to the learners. Many researchers have noticed that botany has been neglected in biology teaching for some time (Anderson, Evertson, & Emmer, 1980; Evertson, Emmer, Sanford, & Clements, 1983; Rennie, 2007), and the US National Research Council (1992) has also identified a decline in plant biology training and research. Consequently, there is no doubt that people's 'plant blindness', a term used by Wandersee and Schussler (2001) to describe the inability to see or notice the plants in one's own environment, is getting worse. With respect to this, the BGEs have the responsibility to engage the public in learning about the plants and their uses, which requires them to have a deeper understanding about botany and fundamental skills in horticulture. In this study, some BGEs showed their expertise in botany and horticulture. For example, Julia told the students many plants' scientific names and Mark explained to his the basic principles of plant taxonomy, whereas the others overlooked this.

Last but not least, the BGE has the responsibility to engage the public with conservation. Promoting biodiversity and growing visitors' environmental awareness through public education programmes is a key working agenda for all the botanic gardens throughout the world (BGCI, 2008; Willison, 2006), and in many botanic gardens, addressing environmental issues has been listed in the BGEs' mission statement. However, through the field observations of the four BGEs' teaching practices in this study, it is noted that only Debbie integrated the elements of environmental education into her guided visits. In her investigation on the collaborative relationship between an English botanic garden and local schools, Vergou (2010) found that BGEs did not address environmental issues to the visiting schoolchildren, because the schoolteachers preferred the visit to be related to learning at school and the National Curriculum. Oikawa (2000) explains that it was because of the BGEs' training background, pointing out that:

> There is a strong tendency for educators to be trained as school teachers ..., but no advanced qualification in natural sciences. That makes them effective at responding to the demands of school groups, but can marginalize education both structurally and philosophically within the organization. (p. 425)

It is apparent that the BGEs should not only think of responding to the demands of schoolteachers but also consider including conservation education, so as to enhance their visitors' understanding about the concept of sustainability.

7.2.2   Implications for Theories

The findings from this study provide further support for the role of mediation, advocated by the sociocultural theory, which involves the practice of scaffolding understanding. The term scaffolding, defined by Wood, Bruner and Ross (1976), refers to a process 'that enables a child or novice to solve a task or achieve a goal that would be beyond his unassisted efforts' (p. 90). They noted that this process requires the adult to manipulate 'those elements of tasks that are initially beyond the learner's capacity, thus permitting him to concentrate upon and complete only those elements that are within his range of competence' (Wood et al., 1976, p. 90). That is, the adult's appropriate assistance can bridge what students already know to arrive at something they do not know. However, providing appropriate scaffolding is a relatively more challenging task for an informal educator than it is for a school-teacher, because the former usually does not have sufficient knowledge about the students' prior experience and understanding (King, 2009). With respect to this, some BGEs in the current study adopted certain strategies, such as assessing students' previous knowledge and checking for consensus during their guided visits, so as to identify the gap between what the students already knew and what new knowledge they could acquire from the botanic garden setting. Moreover, during the interviews the BGEs reported that their communications with the schoolteachers prior to the visit about the students' background and the level of their academic attainment target were helpful for organizing the visits more effectively, because they could use them to ensure that their teaching met the current level of the students' knowledge and ability.

Proponents of the sociocultural theory of learning, on which this study has been based, maintain that learning is constructed through interactions among individuals and emphasizes the importance of discourse in the process. In this regard, as a form of scaffolding, building and maintaining learning conversations between an educator and students can improve the level of understanding. Although previous research has shown that an educator's 'maintain', 'elicit' and 'press' moves when responding to student statements and questions have the potential to further elaborate the student thoughts and extend learning conversations (Brodie, 2009; Edwards & Mercer, 1987; King, 2009; Scott, 1998), there is no evidence in this study to support such findings. As presented in Chap. 5, the proportions of 'maintain', 'elicit' and 'press' utterances used by Simon and Debbie to follow up their student talk were relatively higher than other BGEs; however, chained triadic pattern of discourses (I-R-F-R-F-R-F-) in exchanging ideas between them and their students was not commonly observed. Moreover, it emerged that social and cultural factors, such as language barriers (i.e. where many students in a group speak English as an additional language), can have an impact on the nature of a lesson's discourse.

Another perspective that relates to the sociocultural theory is that gestures should become a more important factor for shaping how a learner appropriates shared knowledge. That is, in the process of developing shared knowledge, much of what we mean is not only partially constructed through linguistic means, such as presenting ideas to the class, sharing individual student's experience with the class, revoicing a student's idea and using 'we' statements, but is also partially constructed through gestures (Edwards & Mercer, 1987; O'Connor & Michaels, 1996; Roth, 2001; Scott, 1998). The evidence from the BGEs' pedagogical moves, especially their metaphorical modelling, as discussed in Chap. 6, has supported this perspective that meaning is manifested not only verbally but also gesturally. Moreover, in some cases, such as when communicating abstract and complex scientific processes, gestural signals may even be superior to those verbally expressed, when explaining an idea or concept, particularly when it is abstract or complex (Lantolf, 2004).

The next implication of relevance to the sociocultural theory is that the promotion of student autonomy is a desired goal, and in particular, researchers have called for students to be allowed more authority over their learning on school excursions to informal settings (Griffin, 2004). In this research, a higher proportion of student volunteering utterances were found in Mark and Julia's observed lessons than in those of the others, which indicates that these two BGEs were more supportive of student autonomy, allowing the students to play a more proactive role in the lesson discourse. Moreover, it is noted that during these two BGEs' guided visits, the student activities mainly involved active-directed and guided explorations, which are aimed at encouraging children to participate. Furthermore, both Mark and Julia in their interviews reported that they requested the schoolteachers to prepare their students prior to the visits. In addition, the students observed in these two BGEs' visits were mainly from independent schools, where they had more experiences of going on school trips and, thus, they were permitted to be more self-regulating during the visits, because there was a strong level of trust. In sum, although autonomy in the learning process for students is a desired goal in botanic garden settings, this has to be framed by well-structured activities being provided by the educators so as to ensure that the learners remain on task.

7.3   Limitations of the Study and Further Research

This study was designed to address three questions: 'What are the structures of the BGEs' guided school visits?' 'How do BGEs interact with visiting schoolchildren in terms of engaging and supporting learning?' 'What pedagogical behaviours can be observed during the guided visit?' To investigate these questions a range of data was collected and a variety of research strategies was used. However, some limitations emerged with regard to the data set and the research methods.

First of all, the size of the sample in this study was relatively small, with only four BGEs being recruited from three participating sites. This was because for a number of reasons, as presented in Chap. 4, access to botanic gardens was problematic. Consequently, the generalizability of the specific roles that BGEs have been or should be adopting is not appropriate. However, because the study was aimed at investigating, in detail, the practices of the BGEs concerned and it emerged that there was sufficient overlap in the jobs that they were performing, I would propose that the findings are still robust enough to be considered as having identified the key functions that BGEs should be engaged in. That is, they not only correctly map out what effective BGEs should be doing but also provide a useful springboard for future research in this field. Nevertheless, it is accepted that with such a small sample, other aspects of BGE practice may have been missed.

The records pertaining to visiting school groups in the three participating botanic gardens showed that there were more school groups in the spring and summer terms than the rest of the year and this information is in line with the findings in other outdoor education research that school field trips were influenced by the weather (Kisiel, 2005; Orion & Hofstein, 1994; Schlossberg, Greene, Phillips, Johnson, & Parker, 2006). In other words, the school education programmes in botanic gardens vary according to the season, and ideally, the data collected for this research should have covered the school visits throughout the whole academic year. However, the visiting school groups willing to participate had planned their school trips only for the summer term.

Although the impression sheets completed by the schoolchildren reflected their visiting experiences and perceptions of the BGE's teaching, more information could have been collected if I had been able to interview them prior to and after the trip. However, as explained in Sect. 4.4.3, the plans to interview students were dropped, because of concerns regarding the potential unwillingness of the participating school groups to comply, and owing to my personal background as an overseas student who speaks English as a foreign language, who has not had much work experience with primary and secondary children, I did not believe I would be considered suitably qualified to be granted permission to do so. Moreover, there may well have been insurmountable security clearances issues involved.

All the BGEs were interviewed and asked to reflect on their teaching practices a few weeks after the observation took place. Although I was able to show them video clips and the transcribed discourse data to jog their memories during these interviews, it would have better if I had been able to speak to them much sooner after their lessons, but this was not logistically feasible. I had thought of asking them to record their impressions of the observed sessions, soon after they had finished, but on reflection decided that this was probably asking too much of them, given their busy work schedule. Consequently, some of their interpretations of what had happened during the visits in question may well have been not as insightful as they could have been with their recollections fresher.

As discussed in Chaps. 5 and 6, this study has highlighted a number of areas on which to focus further research into school educational excursions to botanic gardens, in particular, the pedagogical practice that BGEs should adopt so as to engage and support children's learning. The findings from this study raise a number of questions which could be addressed in further research.

What changes to the BGEs' pedagogical practice would be most effective in improving the level of engagement of students? This would involve an interventionist or action research programme, whereby the BGEs would be provided appropriate professional development training aimed at increasing their understanding of the nature of science. The design of the professional development programme could follow the framework that is presented in this chapter and investigate the difference in their class discourse prior to and after the intervention. By so doing, not only would the framework of pedagogy developed from this study be tested, but also it would provide opportunities for the BGEs involved to reflect on their practice, both individually and collectively.

What is the difference in the pedagogy that BGEs employ for teaching primary and secondary school groups? This study investigated primary school groups on the BGE-guided botanic garden visits, with the exception being Debbie's Year 8 groups, owing to the accessibility of the participants. It appears that there are fewer secondary school visits to botanic gardens than primary ones, even though some BGEs have designed some programmes for GCSE and A-level students. Further research could set out to find out what pedagogy is most appropriate for secondary groups in terms of improving their acquisition knowledge capabilities. Moreover, a comparative study of the teaching strategies used for primary and secondary students in outdoor settings could enrich our understanding of how the BGEs adapt their practice to meet different age groups' needs. It would also reveal areas of strengths or the weaknesses regarding current botanic garden education for visiting school groups.

What kind of activities might support learning from a visit led by a BGE whose objective is to let the children 'see it all'? As recommended earlier in this chapter, the design of education programmes may be divided into two groups, the 'general' visit and the 'focused' visit. For the general visit, the BGE usually has the objective of enabling visiting schoolchildren to see the botanic garden as a whole. The purpose of such an education programme is to enrich children's botanical experience and enhance their interest to learn science. In this sense, it is important to investigate what kind of activities in limited visiting time might support the achievement of that particular education purpose.

To what extent can the experience of a botanic garden visit led by a BGE support positive attitudes towards science and the environment? The ultimate goal of botanic garden education for visiting schoolchildren is to develop their interest in doing science and enhancing environmental awareness. School trip experiences have been shown not only to influence children's cognitive development but also social and affective outcomes (Dillon, Morris, O'Donnell, Reid, & Rickinson et al., 2005; Meredith, Fortner, & Mullins, 1997; Nundy, 1999; Wellington, 1990). For example, Jarvis and Pell (2005), who surveyed the attitude changes of 300 children aged 10 or 11 years old visiting the UK National Space Centre, found that a short school trip experience to an informal science institute can have a long-lasting positive impact. They reported that nearly 20 % of these children showed an increased interest in science and in pursuing a science career several months after the visit. Compared to the visitors' experience to museum settings, little is known about the impact of botanic garden visits in this respect, and a possible avenue for further research is to conduct a longitudinal study examining the impact of such experiences on student perceptions and their future career choices.

References

Allen, S. (2002). Looking for learning in visitor talk: A methodological exploration. In G. Leinhardt, K. Crowley, & K. Knutson (Eds.), *Learning conversations in museums* (pp. 259–303). Mahwah: Lawrence Erlbaum Associates.

Anderson, D., Piscitelli, B., Weier, K., Everett, M., & Tayler, C. (2002). Children's museum experiences: Identifying powerful mediators of learning. *Curator, 45*(3), 213–231.

Anderson, L. M., Evertson, C. M., & Emmer, E. T. (1980). Dimensions in classroom management derived from recent research. *Journal of Curriculum Studies, 12*(4), 343–356.

Ash, D., & Wells, G. (2006). Dialogic inquiry in classroom and museum: Actions, tools and talk. In Z. Bekerman, N. C. Burbules, & D. S. Keller (Eds.), *Learning in places: The informal education reader* (pp. 35–54). New York: Peter Lang.

Bamberger, Y., & Tal, T. (2007). Learning in a personal context: Levels of choice in a free choice learning environment in science and natural history museums. *Science Education, 91*(1), 75–95.

Bedford, L. (2001). Storytelling: The real work of museums. *Curator: The Museum Journal, 44*(1), 27–34.

BGCI. (2008). Mission statement. Retrieved November 27, 2008, from http://www.bgci.org/global/mission/

Brodie, K. (2009). *Teaching mathematical reasoning in secondary school classrooms*. Dordrecht: Springer.

Carson, R. (1998). *The sense of wonder*. New York: HarperCollins.

Cazden, C. B. (2001). *Classroom discourse: The language of teaching and learning* (2nd ed.). Portsmouth: Greenwood Press.

Chin, C. (2007). Teacher questioning in science classrooms: Approaches that stimulate productive thinking. *Journal of Research in Science Teaching, 44*(6), 815–843.

Cornell, J. B. (1989). *Sharing nature with children 2*. Nevada City: Dawn Publications.

Cox-Petersen, A. M., Marsh, D. D., Kisiel, J., & Melber, L. M. (2003). Investigation of guided school tours, student learning, and science reform recommendations at a museum of natural history. *Journal of Research in Science Teaching, 40*(2), 200–218.

Crowley, K., Callanan, M. A., Jipson, J. L., Galco, J., Topping, K., & Shrager, J. (2001). Shared scientific thinking in everyday parent–child activity. *Science Education, 85*(6), 712–732.

Cuevas, P., Lee, O., Hart, J., & Deaktor, R. (2005). Improving science inquiry with elementary students of diverse backgrounds. *Journal of Research in Science Teaching, 42*(3), 337–357.

DeWitt, J., & Hohenstein, J. (2010). School trips and classroom lessons: An investigation into teacher-student talk in two settings. *Journal of Research in Science Teaching, 47*(4), 454–473.

DeWitt, J., & Osborne, J. (2007). Supporting teachers on science-focused school trips: Towards an integrated framework of theory and practices. *International Journal of Science Education, 29*(6), 685–710.

DeWitt, J., & Storksdieck, M. (2008). A short review of school field trips: Key findings from the past and implications for the future. *Visitor Studies, 11*(2), 181–197.

DfES. (2006). *Learning outside the classroom manifesto*. Nottingham: Department for Education and Skills (DfES) Publications.

Dillon, J., Morris, M., O'Donnell, L., Reid, A., Rickinson, M., & Scott, W. (2005). *Engaging and learning with the outdoors: The final report of the outdoor classroom in a rural context action research project*. London: National Foundation for Education Research.

Driver, R. (1989). Students' conceptions and the learning of science. *International Journal of Science Education, 11*(5), 481–490.

Edwards, D., & Mercer, N. (1987). *Common knowledge: The development of understanding in the classroom*. London: Routledge.

Evertson, C. M., Emmer, E. T., Sanford, J. P., & Clements, B. S. (1983). Improving classroom management: An experiment in elementary school classrooms. *The Elementary School Journal, 84*(2), 173–188.

Falk, J. H., & Balling, J. D. (1982). The field trip milieu: Learning and behavior as a function of contextual events. *Journal of Educational Research, 76*(1), 22–28.

Falk, J. H., & Dierking, L. D. (1992). *The museum experience.* Washington, DC: Whaleback.

Gilbert, J., & Priest, M. (1997). Models and discourse: A primary school science class visit to a museum. *Science Education, 81*(6), 749–762.

Grenier, R. S. (2009). The role of learning in the development of expertise in museum docents. *Adult Education Quarterly, 59*(2), 142–157.

Griffin, J. M. (1998). Learning science through practical experiences in museums. *International Journal of Science Education, 20*(6), 655–663.

Griffin, J. M. (2004). Research on students and museums: Looking more closely at the students in school groups. *Science Education, 88*(Suppl. 1), S59–S70.

Griffin, J. M., & Symington, D. (1997). Moving from task-oriented to learning-oriented strategies on school excursions to museums. *Science Education, 81*(6), 763–779.

Hein, G. E. (1998). *Learning in the museum.* New York: Routledge.

Hofstein, A., & Rosenfeld, S. (1996). Bridging the gap between formal and informal science learning. *Studies in Science Education, 28*(1), 87–112.

Jarvis, T., & Pell, A. (2005). Factors influencing elementary school children's attitudes toward science before, during, and after a visit to the UK National Space Centre. *Journal of Research in Science Teaching, 42*(1), 53–83.

Jonassen, D. (2004). Towards a design theory of problem solving. *Educational Technology Research and Development, 48*(4), 68–85.

Kellert, S. R. (2002). Experiencing nature: Affective, cognitive, and evaluating development in children. In P. H. Kahn & S. R. Kellert (Eds.), *Children and nature: Psychological, sociocultural, and evolutionary investigations* (pp. 117–151). Cambridge, MA: The MIT Press.

King, H. (2009). *Supporting natural history enquiry in an informal setting: A study of museum explainer practice (Unpublished doctoral dissertation).* London, UK: King's College London.

Kisiel, J. (2005). Understanding elementary teacher motivations for science trips. *Science Education, 89*(1), 936–955.

Klassen, S. (2009). The construction and analysis of a science story: A proposed methodology. *Science & Education, 18*(3–4), 401–423.

Kuh, L., Ponte, I. C., & Chau, C. (2010). *Play behaviors before and after a natural playground installation in an early childhood setting.* Somerville: Tufts University.

Lantolf, J. P. (2004). Introducing sociocultural theory. In J. P. Lantolf (Ed.), *Sociocultural theory and second language learning* (pp. 1–26). Oxford: Oxford University Press.

Lobato, J., Clarke, D., & Ellis, A. B. (2005). Initiating and eliciting in teaching: A reformulation of learning. *Journal for Research in Mathematics Education, 36*(2), 101–136.

Malone, K. (2007). The bubble-wrap generation: Children growing up in walled gardens. *Environmental Education Research, 13*(4), 513–527.

Martin, M., Brown, S., & Russell, T. (1991). A study of child-adult interaction at a natural history centre. *Studies in Educational Evaluation, 17*(2–3), 355–369.

Matthews, M. R. (1989). History, philosophy, and science teaching: A brief review. *Synthese, 80*(1), 1–7.

Meredith, J. E., Fortner, R. W., & Mullins, G. W. (1997). Model of affective learning for nonformal science education facilities. *Journal of Research in Science Teaching, 34*(8), 805–818.

Mortimer, E. F., & Scott, P. H. (2003). *Meaning making in secondary science classroom.* Maidenhead: Open University Press.

National Research Council, U. S. (1992). *Plant biology research and training for the 21 century.* Washington, DC: National Academy Press.

Negrete, A., & Lartigue, C. (2004). Learning from education to communicate science as a good story. *Endeavour, 28*(3), 120–124.

Nundy, S. (1999). The fieldwork effect: The role and impact of fieldwork in the upper primary school. *International Research in Geographical and Environmental Education, 8*(2), 190–198.

Nystrand, M. (1997). *Opening dialogue: Understanding the dynamics of language and learning in the English classroom.* New York: Teachers College Press.

Nystrand, M., Wu, L. L., Gamoran, A., Zeiser, S., & Long, D. A. (2003). Questions in time investigating the structure and dynamics of unfolding classroom discourse. *Discourse Processes, 35*(2), 135–198.

O'Connor, M. C., & Michaels, S. (1996). Shifting participant frameworks: Orchestrating thinking practices in group discussion. In D. Hicks (Ed.), *Discourse, learning, and schools* (pp. 63–103). Cambridge, UK: Cambridge University Press.

Ohlsson, S. (1996). Learning to do and learning to understand: A lesson and a challenge for cognitive modelling. In P. Reiman & H. Spada (Eds.), *Learning in humans and machines: Towards an interdisciplinary learning science* (pp. 37–62). Oxford, UK: Elsevier.

Oikawa, J. (2000). *Future role of living plant collections in gardens for biodiversity conservation* (Unpublished doctoral dissertation). University of Reading, Reading, UK.

Orion, N., & Hofstein, A. (1994). Factors that influence learning during a scientific field trip in a natural environment. *Journal of Research in Science Teaching, 31*(10), 1097–1119.

Osborne, J. (1998). Constructivism in museums: A response. *Journal of Museum Education, 23*(1), 8–9.

Paris, S. G., Yambor, K. M., & Packard, B. (1998). Hands-on biology: A museum-school-university partnership for enhancing students' interest and learning in science. *The Elementary School Journal, 98*(3), 267–288.

Price, S., & Hein, G. E. (1991). More than a field-trip: Science programmes for elementary school groups at museum. *International Journal of Science Education, 13*(5), 505–519.

Ramey-Gassert, L., & Walberg, H. J. (1994). Re-examining connections: Museums as science learning environments. *Science Education, 78*(4), 345–363.

Rennie, L. J. (2007). Learning science outside of school. In S. K. Abell & N. G. Lederman (Eds.), *Handbook of research on science education* (pp. 125–167). Mahwah: Lawrence Erlbaum.

Rickinson, M., Dillon, J., Teamey, K., Morris, M., Choi, M. Y., Sanders, D., et al. (2004). *A review of research on outdoor learning.* London: National Foundation for Educational Research & King's College London.

Roth, W.-M. (2001). Gestures: Their roles in teaching and learning. *Review of Educational Research, 71*(3), 365–392.

Russell, I. (1990). Visiting a science centre: What's on offer? *Physics Education, 25*(1), 258–262.

Schlossberg, M., Greene, J., Phillips, P. P., Johnson, B., & Parker, B. (2006). School trips Effects of urban form and distance on travel mode. *Journal of the American Planning Association, 72*(3), 337–346.

Scott, P. H. (1998). Teacher talk and meaning making in science classrooms: A Vygotskian analysis and review. *Studies in Science Education, 32*(1), 45–80.

Scott, P. H., Mortimer, E. F., & Aguiar, O. G. (2006). The tension between authoritative and dialogic discourse: A fundamental characteristic of meaning making interactions in high school science lessons. *Science Education, 90*(4), 605–631.

Smithenry, D. W. (2010). Integrating guided inquiry into a traditional chemistry curricular framework. *International Journal of Science Education, 32*(13), 1689–1714.

Stavrova, O., & Urhahne, D. (2010). Modification of a school programme in the Deutsches museum to enhance students' attitudes and understanding. *International Journal of Science Education, 32*(17), 2291–2310.

Stigler, J. W., Gonzales, P. A., Kawanka, T., Knoll, S., & Serrano, A. (1999). *The TIMSS videotape classroom study: Methods and findings from an exploratory research project on eighth-grade mathematics instruction in Germany, Japan, and the United States.* Washington, DC: National Center for Education Statistics, U.S. Department of Education.

Tal, T. (2012). Imitating the family visit: Small-group exploration in an ecological garden. In E. Davidsson & A. Jakobsson (Eds.), *Understanding interactions at science centers and museums: Approaching sociocultural perspectives* (pp. 193–206). Rotterdam: Sense.

Tal, T., & Morag, O. (2007). School visits to natural history museums: Teaching or enriching? *Journal of Research in Science Teaching, 44*(5), 747–769.

Tenenbaum, G., Naidu, S., Jegede, O., & Austin, J. (2001). Constructivist pedagogy in conventional on-campus and distance learning practice: An exploratory investigation. *Learning and Instruction, 11*(2), 87–111.

Tunnicliffe, S. D. (2001). Talking about plants: Comments of primary school groups looking at plant exhibits in a botanical garden. *Journal of Biological Education, 36*(1), 27–34.

Vergou, A. (2010). *An exploration of botanic garden-school collaborations and student learning experiences.* Ph.D. doctoral dissertation, University of Bath, Bath, UK.

Wandersee, J. H., & Schussler, E. E. (2001). Toward a theory of plant blindness. *Plant Science Bulletin, 47*(1), 2–9.

Wellington, J. (1990). Formal and informal learning in science: The role of the interactive science centres. *Physics Education, 25*(5), 247–253.

Wellington, J., & Osborne, J. (2001). *Language and literacy in science education.* Buckingham: Open University Press.

Wilde, M., & Urhahne, D. (2008). Museum learning: A study of motivation and learning achievement. *Journal of Biological Education, 42*(2), 78–83.

Willison, J. (2006). *Education for sustainable development: Guidelines for action in botanic gardens.* Richmond: Botanic Gardens Conservation International.

Windschitl, M. (2002). Framing constructivism in practice as the negotiation of dilemmas: An analysis of the conceptual, pedagogical, cultural, and political challenges facing teachers. *Review of Educational Research, 72*(2), 131–175.

Wood, D., Bruner, J. S., & Ross, G. (1976). The role of tutoring in problem solving. *Journal of Child Psychology and Psychiatry, 17*(2), 89–100.

Zubrowski, B. (2009). *Exploration and meaning making in the learning of science.* Dordrecht: Springer.

Index

© Springer Science+Business Media Singapore 2015　　　　　　　　　　175
J. Zhai, *Teaching Science in Out-of-School Settings*,
DOI 10.1007/978-981-287-591-4

Printed in the United States
By Bookmasters

Printed in the United States
By Bookmasters